E-Mobility and Related Clean Technologies from an Empirical Corporate Finance Perspective

Finanzmärkte und Klimawandel

Herausgegeben von
Dirk Schiereck und Paschen von Flotow

Band 6

PL ACADEMIC RESEARCH

Christian Stephan Babl

E-Mobility and Related Clean Technologies from an Empirical Corporate Finance Perspective

State of Economic Research, Sourcing Risks, and Capital Market Perception

PL ACADEMIC RESEARCH

Bibliographic Information published by the Deutsche Nationalbibliothek
The Deutsche Nationalbibliothek lists this publication in the Deutsche Nationalbibliografie; detailed bibliographic data is available in the internet at http://dnb.d-nb.de.

Zugl.: Darmstadt, Techn. Univ., Diss., 2014
Schriftenreihe Finanzmärkte und Klimawandel; Bd. 6

Library of Congress Cataloging-in-Publication Data
Babl, Christian Stephan, 1979-
E-mobility and related clean technologies from an empirical corporate finance perspective : state of economic research, sourcing risks, and capital market perception / Christian Stephan Babl.
pages cm. — (Finanzmärkte und Klimawandel ; Band 6)
ISBN 978-3-631-66139-0
1. Clean energy industries—Finance. 2. Green technology—Economic aspects.
3. Electric automobiles—Economic aspects. I. Title.
HD9502.5.C542B33 2015
658.5'14—dc23

2015018823

D 17
ISSN 2190-3069
ISBN 978-3-631-66139-0 (Print)
E-ISBN 978-3-653-05708-9 (E-Book)
DOI 10.3726/978-3-653-05708-9

PL Academic Research is an Imprint of Peter Lang GmbH.

Peter Lang – Frankfurt am Main · Bern · Bruxelles · New York · Oxford · Warszawa · Wien

This publication has been peer reviewed.

www.peterlang.com

Selene Emilie

Geleitwort

Kaum einem anderen Bereich der sektororientierten Betriebswirtschaftslehre ist in den letzten Jahren so viel Aufmerksamkeit zuteil geworden wie dem Bereich, der inzwischen gerne als ‚Renewables' ober ‚Cleantech' bezeichnet wird. Mit dem deutlich zugenommenen und weiter hohen Forschungsinteresse in diesem Bereich ist auch die Anzahl an akademischen Publikationen in allgemeinen betriebswirtschaftlichen Zeitschriften wie auch in spezifisch ausgerichteten Fachjournalen stark gewachsen, und der Stand der Forschung wird zusehends unübersichtlicher. Es ist bzw. es war an der Zeit, dass sich jemand daran machte, Ordnung und Transparenz zu schaffen. Es ist die Sisyphusarbeit der zu erheblichen Teilen auch händischen Erfassungen von vielen hundert Aufsätzen, die vielleicht die rückblickend am wenigsten gewürdigte Leistung in der vorliegenden Dissertationsschrift ausmacht. Die hier vorgestellten Literaturauswertungen schaffen aber eine Klarheit für nachfolgende Forschergenerationen, die kaum überschätzt werden kann. Sie eröffnen zudem mit der Darstellung des Vorhandenen auch den Blick auf das, was noch fehlt, auf die Forschungslücken, die zu befüllen das beständige Ziel des wissenschaftlichen Erkenntnisgewinns darstellt. Und so möchte man nur allen an Finanzierungsfragen der Energiewende interessierten Forschern zurufen: Lest hier, was es gibt, und erforscht, was es nicht gibt! Allerdings wäre es mehr als ungerecht, die Forschungsleistungen von Herrn Babl auf die Literaturauswertungen zu reduzieren. Die politisch gewünschte und von breiten Bevölkerungsteilen getragene Energiewende soll sich insbesondere auch im Individualverkehr mit der Elektromobilität durchsetzen. Hier bauen viele Hersteller auf Joint Ventures, deren Erfolg und Erfolgscharakteristika bislang kaum bekannt sind. Abgesehen von der Organisationsform stellt sich für diese Unternehmen bei kritischen Inputfaktoren, den Seltenen Erden, die Frage, inwieweit Beschaffungspreisrisiken über Finanzmärkte abgesichert werden können. Die vorliegende Arbeit nimmt sich diesen bedeutenden Forschungslücken mit viel Liebe zum Detail und höchster Sorgfalt an. Im Ergebnis wird ein empirisch fundierter Kenntnisstand erreicht, auf dessen Basis sich klare Handlungsempfehlungen für die Industriepraxis ableiten lassen. Zudem finden sich auch für wirtschaftspolitische Entscheidungsträger wichtige Informationen zum Verständnis und zur Bewertung der für die Energiewende so kritischen Fragen. Herr Babl kann die selbst gesetzten Ziele in seiner Dissertationsschrift bestens erfüllen. Die Arbeit enthält viele hoch interessante Resultate und ist so geschrieben, dass es den Lesern große Freude machen wird, sie Seite für Seite bis zum Ende zu studieren. Ich wünsche der Arbeit eine weite Verbreitung.

Professor Dr. Dirk Schiereck

Preface and Acknowledgement

The following pages summarize my work over more than three years. It is my doctoral thesis on financial issues pertaining to the Clean technology (Cleantech) industry, and especially the automotive e-mobility industry. It represents only a small scientific extract of the much bigger "Climate Change Financial Markets and Innovation (CFI)" project which has been conducted by Sustainable Business Institute (Sustainable Business Institute (SBI)) and my alma mater Technische Universität Darmstadt.

Within the CFI project, Technische Universität Darmstadt and our corporate finance department under the leadership of Prof. Dr. Dirk Schiereck has pursued the goal of outlining the financial needs of the Cleantech industry as complete as possible. During our time, my colleagues and I have created many works including studies on venture capital, project finance, debt financing, Mergers & Acquisitions (M&A) or Initial Public Offering (IPO) issues. Besides, I had the opportunity to position myself within the project. I originally come from the automotive industry and it was no coincidence that all issues which dealt with the desired and difficult transition to an electric-driven society landed on my desk. From an empirical finance perspective, it was not an initially rewarding task, since the firms with a pure sustainable mobility focus are rare and many of them are still struggling and longing for market breakthrough. Hence, many small firms were not accessible for analysis as they are not publicly listed on a stock exchange. However, I soon realized that the industry was facing many specific problems which could be identified and partly addressed by empirical financial methods. Especially financial (price) risks and issues which come in line with a transition of supply chains have not really been addressed yet. If the proposed transition to a greater sustainble economy should be more than a catchphrase, it is in my opinion absolutely necessary to rethink all proposed sustainable technologies from an engineering, an environmental and an economic perspective. For Germany as a key manufacturer of automotive products, this insight is even more important. This dissertation may help to address issues of the economic perspective. However, I am still very skeptical whether the transition to e-mobility will succeed in the way politicians imagine. To my mind, it is more likely that the ongoing transformation from mechanical to electric engineering techniques will eventually go on without the help of governmental interventions. While the government might be useful in setting necessary framework conditions and supporting Research & Development (R&D) activities, it should be the demand of the market that finally decides the future and the best solutions.

Before closing this section, I would like to express my gratitude to all the wonderful people I met. To begin with, I would like to thank Prof. Dr. Dirk Schiereck, my doctoral mentor, who was always open to my ideas and gave me good pieces of advice. His kind personality and fine sense of humor have helped me a lot. Many, many thanks also go out to my institute colleagues: Anit Deb, Christoph Ettenhuber, Christian Happ, Achim Himmelmann, Anna Hinrichsen, Florian Kiesel, Sascha Kolaric, Daniel Maul, Steffen Meinshausen, Thomas Poeppe, Malte Raudszus, Julian Trillig and Christopher Welkoborsky. Each one of you has been fantastic and we had a great, almost legendary, time together.

It might look rather unusual, but I would also like to thank the late Hermann Hesse for his wonderful novels and writings. I have read all of them during my time here in Darmstadt and they have given me the necessary balance for heart and soul, even though I feel I am light years away from becoming an enlightened person like his 'Siddhartha'.

And of course I want to express my heartfelt gratitude to my family. First of all, I would like to thank my wife Sonja for her incredible patience. I still wonder how someone could have the ability to put up with me for such a long time, especially my coming home late at night together with my mate and colleague Anit. I need also to thank my mother for being a gorgeous babysitter and taking great care to ensure that I had sufficient time. Last but not least, I want to thank my little daughter Selene Emilie. Since the day she was born, there has never been a day without a smile on my face.

My dissertation is dedicated to you, my dear.

Contents

List of Figures

List of Tables

Acronyms

ACA	Author Co-citation Analysis 15, 33, 41
APT	Arbitrage Pricing Theory 56
CAAR	Cumulated Average Abnormal Return 76
CAR	Cumulative Abnormal Returns 75
CCS	Carbon Capture and Storage 21, 31, 34, 36, 42
CDF	Cumulative Distribution Function 60
CFC	Chlorofluorocarbon 39
Cleantech	Clean technology VII, IX, 4, 6, 7, 9, 31, 32, 33, 35, 39, 41, 42, 45, 85
DCA	Document Co-citation Analysis 15, 33
E-mobility	Electric mobility 1, 2, 4, 5, 7, 31, 40, 63, 67, 68, 69, 71, 72, 82, 84, 85, 86, 87
EBITDA	Earnings Before Interests, Taxes, Depreciation and Amortization 82
EM-Algorithm	Expectation-Maximization-Algorithm 55
EUM	Expected Utility Maximization 62
EV	Electric vehicle 1, 2, 3, 6, 69, 71, 75, 85
FOB	Free On Board 50, 64
GDP	Gross Domestic Product 1
IPO	Initial Public Offering IX, 4, 8, 68
ISI	Institute for Scientific Information 10
M&A	Mergers & Acquisitions IX, 4, 8, 67, 81
MCMC	Markov Chain Monte Carlo 54
OEM	Original Equipment Manufacturer 5, 8, 67, 84, 86, 87
OLS	Ordinary Least Squares 75
OPEC	Organization of the Petroleum Exporting Countries 49
OTC	Over The Counter 64
PHEV	Plug-in Hybrid Electric Vehicle 1, 2, 3

R&D	Research & Development IX, XVI, 1, 3, 6, 69, 70, 71, 82, 83, 84, 86
REE	Rare Earth Element 45, 47, 49, 63, 64, 65, 85, 86
REO	Rare Earth Oxide 47, 51
SBI	Sustainable Business Institute IX
SCI	Science Citation Index 15, 32
SEO	Seasoned Equity Offering 68
SME	Small and Medium-sized Enterprise 69
ZfU	Zeitschrift für Umweltpolitik & Umweltrecht 18, 21, 29, 34

List of Symbols

A	Adjacency matrix
a_{ii}	Element a_{ii} of adjacency matrix A
α	Angle between two vectors
$\hat{\alpha}$	Regression estimate for intersept
β	Average expected drift β of the observed process
$\hat{\beta}$	Regression estimate for slope
β_Q	Expected drift of the Poisson process in the world of a risk-neutral investor
C	Price of a European Call
$CAAR_{t_1,t_2}$	Cumulated Average Abnormal Return for all observations in event window t_1, t_2
CAR_{i,t_1,t_2}	Cumulated Abnormal Return of firm i in event window t_1, t_2
CC	Collaboration coefficient
CI	Collaboration index
$\mathbb{C}_i$	Set of citation counts i
$C\ it$	Citation counts
$\mathbb{C}_j$	Set of citation counts j
Coc	Co-citation matrix
$CoCit_{ij}$	CoCit-Score of element i, j
d	Amount of change in the case of a downward movement
$\hat{\delta}$	Regression estimate for slope
$G(\phi)T$	Cumulated gains of trading strategy ϕ until time T
$\hat{\gamma}$	Regression estimate for intersept
H	Contingent claims
h	Infinitesimal time h
J	Greatest number of authors per document
j	Number of authors
J_{ij}	Jaccard index matrix of element i, j
K	Strike price K
L	Constant initial costs of an investment
λ	Intensity of Poisson process

$\hat{\lambda}_{ML}$	Maximum likelihood estimator for intensities lambda
$\lambda_{m\mathbb{Q}}$	Intensity rate of the Poisson process m in the world of a risk-neutral investor
$\lambda_{\mathbb{Q}}$	Intensity rate of a Poisson process in the world of a risk-neutral investor
$m_{\pi}(z)$	Moment generating function for the compound Poisson process
μ_{st}	Drift parameter μ at a certain state
N	Number of observations
N_P	Total number of publications
n_j	Number of j-authored publications
$O(h)$	Asymptotic order symbol for infinitesimal time interval h
$\mathbb{P}$	Probabilities of the real probability space
ϕ	Investment portfolio, self-financing trade strategy
p_i	Publication count in journal i
P_{ij}	Transition probability matrix
$\hat{P}_{ij}$	Estimation of the transition probability matrix
π_m	Poisson counting process that belongs to decomposed jumps y_m
π_t^Y	Compound Poisson process at time t with n jumps of size Y
$\pi_{t,\lambda}$	Homogenous Poisson process at time t with intensity lambda
π_t	Poisson counting process at time t
$p_{\mathbb{Q}ym}$	Probability distribution of jump sizes in the world of a risk-neutral investor
ψ	Poisson Cumulative Distribution Function (CDF)
Pu	Weighted publication score
$p(y_m)$	Probability distribution of jumps with jump size y_m
$\mathbb{Q}$	Probabilities in the probability space of a risk-neutral investor
r	Return of a risk-free bond, interest rate
R_0	Spot price of rare earth element at time 0
R	Spot price of rare earth element, R(t)
$Re_{Index,t}$	Return of the market index at time t
$Re_{i,t}$	Return of firm i at time t

S_{ij}	Salton index matrix with elements i, j
σ	Volatility of a diffusion process, standard deviation
T	End of timeline, maturity
t	Time
U	Utility function U
u	Amount of change in the case of an upward movement
$V_t(\phi)$	Value of the self financing portfolio at time t
w_i	Weight of journal i
wi_1	Time interval prior to event date t0
wi_2	Time interval after event date t0
X	Regressor matrix
ξ	Error term
$X(t)$	Counting process, number of events (jumps)
Y	Jump of size Y
y_m	m jumps with equal jump size y
$Z(t)$	Radon-Nikodym derivative of the compound Poisson process

1. Introduction

"You and I come by road or by rail, but economists travel by infrastructure."

—Margaret Thatcher

As early as 1900, the majority of vehicles in the US were steam or electric powered. Diesel- or gasoline-powered vehicles were still rare. Since then, individual mobility has changed drastically, and combustion engines have become the leading technology of the 20th century. However, the negative environmental impact of conventional drive trains has led to changes in thinking. More stringent regulations on exhaust emission standards, rising fuel taxes, or carbon allowance prices have put political pressure on the automotive industry. Carmakers react by rethinking conventional combustion engine concepts in favor of technological concepts with less fuel consumption, carbon reduction, and more eco-friendliness. In the long run, it is the goal not only of engineers but the public to establish private transport solutions at zero emissions (which would also mean electricity generated by renewable energy sources) and competitive costs. For achieving the goal, the long-known idea of electric vehicles has re-emerged and engineers have tried to take advantage of the very efficient (in terms of energy conversion) electric motor technology. In Germany, it is the goal of the Federal Government to develop the market for electric vehicles into a leading market with 100.000 units sold by 2014 and approximately 1 million units by 2020. The figures include vehicle types from Plug-in Hybrid Electric Vehicles (PHEVs) to pure Electric vehicles (EVs)—hybrids are excluded from consideration (Nationale Plattform Elektromobilität, 2012). However, many aspects of the technology (e. g. batteries, power electronics, lightweight construction, standardization) cannot be described as mature and lack marketability. Therefore, national R&D funding for Electric mobility (E-mobility) plays an important role besides private-sector R&D initiatives. The German Federal Government alone funds E-mobility with EUR 1.5 billion or approximately 0.06% of Gross Domestic Product (GDP) p. a (Die Bundesregierung, 2011). Worldwide funding levels of leading states are comparable, with the exception of China. For China estimations calculate a R&D support of approximately EUR 7.5 billion or 0.122% of GDP. Despite significant funding of some leading nations, current sale figures remain at a low level. Fig. 1.1 summarizes current international sales figures of the leading nations based on Bernhart et al. (2013). Market shares are well below 1% in all leading nations and do not indicate a rapid breakthrough of the technology. In absolute figures, however, the US can be seen as an exception with relatively good absolute figures and considerable growth rates. Besides technological risks and political interference, market participants also face several other strategic challenges on the way to E-mobility. For example, Heymann et al. (2013) spot mainly five key challenges for the large-scaled introduction of E-mobility concepts:

- Raw material/commodity needs (Old import dependencies will be replaced by others and may cause new problems.)
- Battery technology (Still the crucial part of the technology that needs further improvements.)
- Change of supply chains (Asian electricity device dominance and the possible impact on local industry market shares.)
- Electricity supply (How should the necessary infrastructure be realized on time?)
- Climate footprint (Will the technology really help to reduce climate-relevant emissions?)

Figure 1.1.: EVs and PHEVs: Worldwide units sold and market shares from Q3 2012 to Q2 2013. Data from Bernhart et al. (2013).

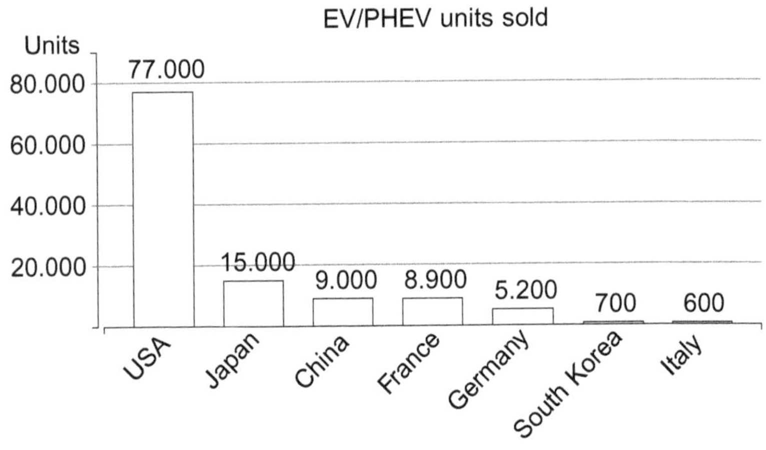

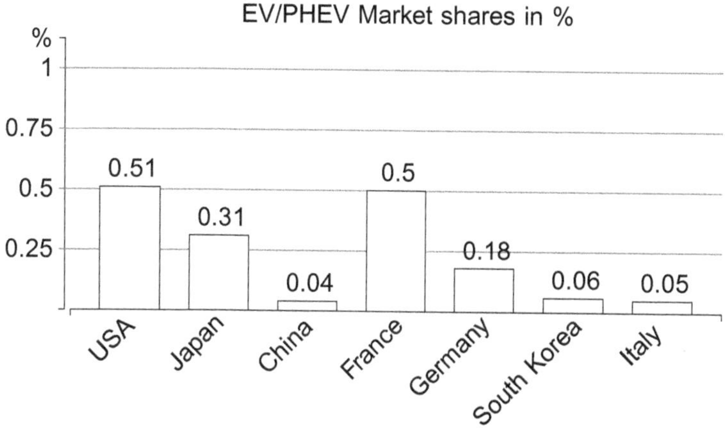

All of the abovementioned challenges reflect the uncertainty about the future development of the technology and pose the strategic question as to how to deal with associated uncertainties and risks. The question does not only affect private automobile manufacturers and their suppliers, but also the aforementioned public sector which needs a long-term planning horizon for infrastructure, education, and financial R&D support. The answer to the question leads to a complicated strategic management task for the private sector and a probably even more difficult planning task for the public sector. The convoluted character of the problem and its proximity to environmental economics clearly imply that solutions will be in fact trans-disciplinary and may therefore combine knowledge from several scientific disciplines such as ecology, engineering, and economy.

However, a closer look at the challenges makes it clear that especially financial economists could strongly contribute to the ongoing discussion and to a partial assessment of the situation. To underline this statement, I want to consider the abovementioned items more in detail. If I consider the proclaimed commodity needs for instance, it becomes clear that the management of supply chain risks is a domain of financial intermediaries. They usually substantially reduce counterpart risk by guaranteeing supply and may help customers to hedge against considerable price risks. While economists know that this has functioned well for the old "oil-based economy" it remains unclear whether the same applies for newly needed material substitutes. In this case, the application of financial methods could be particularly interesting in order to evaluate the sourcing risks that come in line with recently needed substances.

The second and third challenges focus on necessary technological development and the possible change of supply chains. Here again financial methodology and observing capital markets could be beneficial for the evaluation of technological change. Market innovations usually come in line with new capital needs and a "race" for innovation leadership and knowledge. Both can be analyzed by capital market observation - e. g. IPO's, M&A activity or cooperation between the listed firms. Proff and Proff (2013) were the first to mention E-mobility in the context of capital market evaluation and claim that E-mobility still has not played a role yet. Their assessment is supported by reading analysts reports which did not contain many aspects of E-mobility for evaluation. Despite their possible high relevance, concrete capital market studies are still lacking when it comes to capital market evaluation of E-mobility-related technologies.

Viewing in terms of electricity supply and necessary infrastructure, I also find financial aspects which can rather be assigned to classic public finance or project finance. Experience from the German "Energiewende" suggests that immense means would be needed to completely transform our fossil-driven economies into sustainable ones. While this topic does not form part of this dissertation, which rather focuses on an empirical corporate finance analysis, it provides a lot of material for additional scientific evaluation.

The last item on the list is actually determined by the extent of global warming, the actual social cost of carbon emissions and the external effects of fossil fuel consumption. The item shows how the considered technologies enter the sphere of sustainable economics and therefore must be seen in a long-term future with high uncertainties and complexities (Baumgärtner and Quaas, 2010). By promising a reduction of external effects, the technologies are also sometimes referred to as "Cleantech" or "sustainable technology" and may therefore be taken as a justification for governmental interference. The topic may have also led to strong financial implications, for example when financial economists try to calculate the impact of climate change and the social costs of carbon which may, in turn be relevant for the demand of technological substitutes (e. g. Weitzman, 2013). As the last point cannot be directly assigned to an empirical consideration of the market, it plays a rather indirect role in this dissertation, namely when analyzing relevant literature on closely related clean technologies.

Summing up this brief but not exhaustive consideration, it may be stated that many challenges of E-mobility have also to be addressed from a financial market perspective. It is the goal of this dissertation to make use of this insight by writing an assessment for E-mobility and related sustainable technologies from an empirical corporate finance point of view.

The focus on "empirical" methods helps to concentrate on available capital market data which usually merits particular attention in finance. Besides the fact that capital market data can be easily observed, this might come from the fact that capital markets usually price available information very efficiently (Fama, 1970, 1998).[1] Applied to the case of the emerging E-mobility sector, the information efficiency hypothesis of Fama (1970) suggests that the capital market should have already correctly priced all known information on future technologies—a factor which might be exploited by appropriate financial methods. In practical terms, a positive capital market evaluation would mean an important precondition for raising capital which is needed for changing supply chains and the presence in global growth markets (Proff and Proff, 2013). While empirical finance techniques may help to answer the question of whether E-mobility considerations play a role in stock evaluation of automotive Original Equipment Manufacturer (OEM)s and their suppliers, they may also be used for the evaluation of other relevant (supply) markets. Coming back to the issue of changing supply chains and changing commodity needs in the automotive industry, it is only natural to analyze the liquidity and hedging capability of needed commodity markets. The idea of analyzing associated and relevant market areas of E-mobility is realized in chapters 3 and 4 of this dissertation and can be seen as the scientific core. Besides, the work includes a concise review of relevant environmental business and economic literature in order to analyze the current state of research. The next section describes relevant basic definitions and the scope of this dissertation.

1.1. Basic definitions and scope

To begin with, I start with some short definitions of E-mobility and relevant technologies.

E-mobility is a central term of this dissertation and defines vehicles which make use of an electric drive train and receive the energy needed for the electric drive mostly from electricity networks (BMU, 2014).[2] In this definition, it does not make a difference of whether a certain proportion is still driven by combustion engines, or in other words, the definition comprises pure electric vehicle concepts and different types of hybrid cars (Yay, 2010). A relevant technology for E-mobility in this dissertation is a technology that is crucial for the realization

1 The scientific dispute on weaker or stronger forms of the information efficiency hypothesis are of less interest in this introduction.

2 Slightly different definition from original text in BMU (2014): "Elektromobilität im Sinne der Bundesregierung umfasst all jene Fahrzeuge, die von einem Elektromotor angetrieben werden und ihre Energie überwiegend aus dem Stromnetz beziehen, also extern aufladbar sind."

of EVs. The technology term defined here is close related to the terms "clean" or "green technology" and "sustainable technology". It is worth comparing the terms as a key characteristic of the technologies considered here can be clarified. Clean or green technologies are common terms for technologies that help to reduce carbon emissions, raise energy efficiency levels or avoid environmental pollution in general (The Cleantech Group, 2014). Since "Cleantech" is not a scientifically defined term, a clear classification or differentiation of certain technologies seems to be difficult. Instead, the term "sustainable technology" applied by Radke (1996) is more scientifically established. Whereas definitions for "sustainability" exist (Marshall and Toffel, 2005), definitions for "clean" or "green" do not. Due to the lack of a clear scientific definition for the term "Cleantech" experts in the field use the term intuitively when describing a new generation of technologies that solve or help to reduce environmental problems. From an economic point of view, it becomes clear that the technologies described here may be used for the internalization of external effects. Consequently, the price of the considered technologies might differ from "normal technologies" which usually do not take economic externality costs into consideration. Hence, the value of such a "sustainable" technology might strongly depend from externalities (e. g. climate change). As described above, this characteristic is not only of economic interest but can also be seen as a key justification for governmental intervention and the demand for governmental stimulus. The result is that many areas of relevant literature (see also below in Chapter 2) can be assigned to the area of economic and environmental policies and associated political risks which may in turn become a financial risk for investors.

Besides these first definitions of the investigated technologies, I would like to briefly define the scope of this dissertation.

The first part of the dissertation comprises an extensive and thorough literature review and bibliometric evaluation of environmental economics and Cleantech related research in Germany. Especially German scientific output has been selected as an apt research object, as the country had a very early commitment to Renewable Energies and Cleantech applications in conjunction with governmental stimulus and R&D support. The topic fills a gap in current research as the environmental and economic impacts of the latest developments and governmental incentive schemes have not yet been systematically analyzed.[3] The

3 Chapter 2 is an update and summary of previous publications which originated from my work on the doctoral thesis (Babl et al., 2014; Babl, 2014). The publications suggest that a relevant research gap has been identified.

lack of an overall scientific evaluation is remarkable as the interdependence between governmental policies and the desired technological development seems to be crucial for the development of Cleantech related applications and market success.

The second part of the dissertation is based on the results of the literature review and develops an analysis for the E-mobility industry with empirical corporate finance methods. Thus, it is a main goal to make use of available capital market data in order to derive clear scientific statements for the target industry. The requirement for available capital market data reduces the scope of the analysis, thereby excluding other relevant areas of financing E-mobility such as infrastructure finance or public funding (e.g. Brauner, 2008; Leitinger and Litzlbauer, 2011; Dudenhöffer et al., 2012). The term "corporate" makes clear that the work is mainly written from a firm's point of view with a clear business perspective. However, many results also remain valid from a political or macro economic perspective and might therefore be relevant for other market participants involved. Aside from the literature review, the current work is mainly written from a German (as well as European) perspective. This can be seen in the consideration of commodity shortages in Chapter 3 which might be utterly different in other nation states and especially China (see Chapter 3). While the location may play an important role for strategic management decisions, this restriction of scope should not limit the generality of the results.

Before starting with the analysis, the next section allows for the necessary overview by giving a short outline of the dissertation.

1.2. Outline

The dissertation consists of an introductory part, three main parts, and a conclusion. Basically, each chapter represents a separate section with self-contained subjects but the sections are also built on each other (see Figure 1.2).

The actual dissertation begins with a bibliometric literature review of all relevant German speaking research communities in Chapter 2. The analysis consists of a search sensitive part with the goal of detecting all relevant research in the delineated research area. Furthermore, it provides an overview of those publications which had the highest impact on progress and researchers. The chapter summarizes the current state of research and helps to identify gaps in research.

Chapter 3 begins filling such a gap by analyzing the sourcing situation of a very important commodity class for clean tech and e-mobility applications: the rare earth metal market. It mainly examines possible (pricing) risks and therefore hedging possibilities for market participants. In order to realize this, theoretic derivative pricing approaches are systematically applied and deviations from standard theory in terms of classic assumptions are examined.

The capital market perspective on e-mobility and related technologies is provided in Chapter 4. As classic M&A or IPO activity is still infrequently observed in the emerging sector, the impact of cooperations, especially between established OEMs and promising small firms with innovative technologies, is analyzed. Besides statements on the capital market evaluation of the industry, the analysis also provides a first insight into the main evaluation determinants.

The dissertation is rounded off by conclusive remarks and an outlook in the last chapter. It provides the synthesis of all illuminated aspects and summarizes the main results. Furthermore, it contains a prudent valuation and interpretation for future developments from an empirical finance point of view.

Figure 1.2.: Chapter overview. The dissertation consists of three main parts (chapter 2-4), an introduction (Chapter 1), and a synthesis in the form of conclusive remarks and an outlook (Chapter 5).

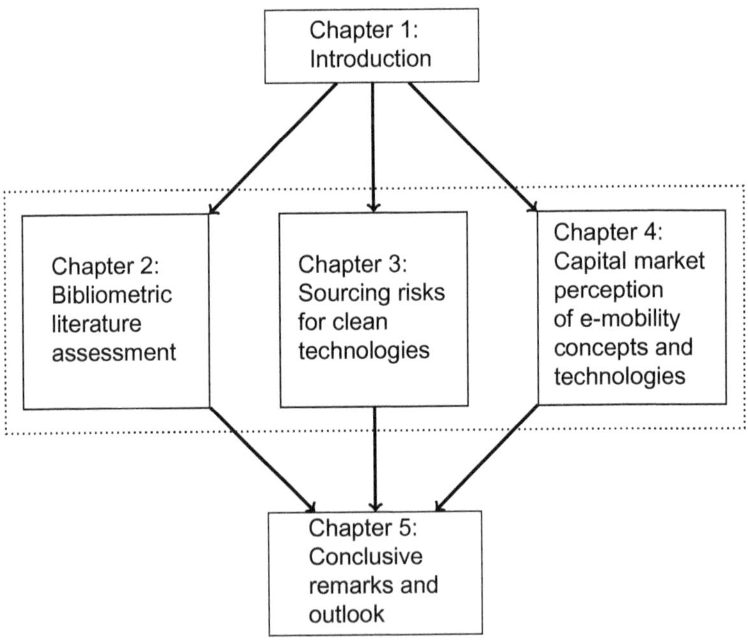

2. Relevant German economic literature - a bibliometric assessment

"To know what you know and what you do not know, that is true knowledge."
—Confucius

The goal of this chapter is to evaluate the latest developments of relevant academic German literature with the help of bibliometric methods. Therefore, the analysis is divided into two parts with different perspectives and scopes. The first part, in section 2.1, deals with the question of how environmental economics and business literature can be characterized as a field of research and what kind of publications have had the most impact. It is a rather abstract but complete view on a complex field of environmental economics research.

The second part deals with the question of whether clean technology-related applications have explicitly been addressed in literature. The methods need to be adopted here, as the question of academic impact plays a tangential role. Instead, rather sensitive methods are applied in order to detect all Cleantech-related literature. The analysis can be found in section 2.2.

2.1. Environmental economics and business literature: An impact analysis

Over the past 40 years, environmental economics as a field of research has become increasingly relevant. Allocation problems and costs associated with pollution, dwindling resources and climate change make instantaneously clear that environmental and ecological problems go hand in hand with economic issues and prosperity. Therefore, the combination of economic and environmental research issues have developed to a mature research area with a wide range of scientific publications media and individual research foci. Especially in the US, the research field has attracted an increasing amount of attention over time and several bibliometric studies have analyzed its development and impact. However, an evaluative bibliometric overview of German speaking research communities is still missing. The goal of this section is to fill this gap with the help of bibliometric methods and to give an overview of the most influential publications, most productive authors and community networks in the field of environmental economics.

While environmental (resource) economics has its origins in neoclassical economics and asks for cost minimal externalities and optimal resource allocation, ecological economics refers to the concept of sustainable development (van den Bergh, 2001). Despite the fact that both approaches can employ substantially different methods, both still address very similar problems at the intersections of economics and ecology. Consequently, the overlap between both disciplines is extensive and the differences give less concern in this dissertation. This is due to the selection criteria of the documents not being motivated by the aforementioned distinctions but by the use of economic and business oriented journal rankings (VHB-Jourqual). Consequently, both terms are treated equivalently, bearing in mind that the difference may yet still be significant.

A more concise view of the literature reveals that the question of structural development and the impact of the field have become increasingly important during the last decade. Costanza et al. (2004) have extracted the most influential publications in the area of ecological economics through a citation analysis. They evaluated Institute for Scientific Information (ISI) data and publications from ecological economics (EE) from 1990-2003 and present publication and related citation ranking lists. Ma and Stern (2006) extend the work of Costanza et al. (2004) and focus on the most influential journal articles and possible differences between environmental and ecological economics between 1994 and 2003. Besides elaborate journal and publication rankings, a distinct analysis of the differences between environmental and ecological economics is provided (in terms of bibliometric measures). Overall, they find the distinction mainly in individual articles and the research focus of the respective journals. Castro e Silva and Teixeira (2011) focus solely on ecological economics and ask whether the discipline has become a post-normal science. With the help of bibliometric methods they evaluated 2,533 articles in the journal Ecological Economics from 1989 to 2009. The results underline that ecological problems have become extremely complex, transdisciplinary, and far-reaching and may therefore not resemble problems of "normal science" in a sense that the interdependence of the separated disciplines becomes more important (Kuhn, 1962; Castro e Silva and Teixeira, 2011).

Before starting with the analysis, a short overview for the following subsections should be provided here. Section 2.1.1 discusses the bibliometric methods applied. It differentiates between descriptive and evaluative oriented approaches. Section 2.1.2 describes the data set and methodical issues which arise with the data collection efforts. In section 2.1.3 the main results are presented. The last section 2.1.4 summarizes the results and refers to the next section with a rather sensitive search approach.

2.1.1. Methods

In this chapter, the goal is to use bibliometric methods in order to evaluate scientific literature and scholarly impact in the area of environmental economics.

Bibliometric methods make quantitative use of metadata that arises with every publication (McCarthy, 1982; Broadus, 1987). Metadata in bibliometrics may cover many categories, such as author names, labels or descriptive information. The choice of data strongly depends on the intended research purpose and type of bibliometric study. According to Potter (1988), bibliometrics can be devided into two main categories: descriptive and evaluative studies. Descriptive studies focus on the productive output of a body of literature, usually by enumerating authors, geographic locations, publication years, or the number of articles published. In contrast, evaluative studies rely on reference and citation data and question how publications are interconnected and used (Potter, 1988; Osareh, 1996). In this chapter, I follow this classification, applying both types of evaluations and discussing them in the following two subsections. Therefore, the first subsection covers and summarizes the basic methods for finding the most influential publications based on descriptive meta data, such as citation counts and the total number of publications. The second subsection then discusses how community detection methods can be modified in order to reveal the most influential publications in a field and their interconnections.

Descriptive methods

This section provides a brief overview of the descriptive methods for extracting the most influential publications and authors of a research field. But first of all, the data requirements for evaluation should be defined. For this reason, some bibliometric indicators which reflect facets of "influence" and scientific collaboration will be introduced and data requirements are derived subsequently. A good overview of the most popular bibliometric indicators can be found in Okubo (1997); Verbeek et al. (2002). The main indicators are structured according to Havemann (2009).

- Document impact
  Within bibliometric metadata, citations still merit a high degree of attention as a quantitative indicator for scientific impact. It is commonly assumed that many citations go in line with high attention and may therefore indicate relevancy. However, besides technical problems, there are several methodological pitfalls which need to be addressed thoroughly.[4] First of all, a proper counting method needs to be specified. The most straightforward method is to simply add up the number of citations for each selected document. However, measuring impact solely by citation counts may easily lead to misinterpretations

4 For technical issues of data collection see data section.

and distorted results. For example, a problem occurs as soon as documents with several authors are cited. While fractional counting may be a solution to measure the impact of each author, the impact of a document is rather nonfractional and can only be counted in integers. As Van Raan (2005) states, it could also be very important to differentiate between different types of documents: Among others, one can distinguish between ordinary "articles", "research notes", "errata" or "reviews". Van Raan (2005) argues that all of the different types of documents may strongly influence the citation behavior. Besides common articles, purely introductiory texts which may cite or be cited more often, are excluded from ranking positions. Another issue may be caused by the choice of language (Van Raan, 2005). As long as the hand collected data set is evaluated, the influence of linguistic problems can be left aside, as I concentrate solely on German research communities. Nonetheless, the influence of German written articles in terms of impact is expected to be (very) low in comparison with Anglo-American publications that appeal to a wider spectrum of researchers.

- Authorship productivity
  While impact measures may indicate the most cited and most noticed documents (or journals) of a field, they do not tell much about the productivity of a single author within a research community. Just as with citation counts, publication counts may simply be added up as an indication for productive output (Okubo, 1997). The first who applied this method was Lotka (1926). Lotka (1926) intended to measure the contribution of each author to a research community which led to today's "Lotka's law". However, the evaluation of pure publication numbers is comparable to the situation with citation counts, and one must be aware that there may be several pitfalls and problems which will be discussed shortly. Just as with citation counts, collaboration and co-authorship of n authors is an issue. Especially in the case of total publication numbers, bibliometricians usually apply fractional counts, such as 1/n for each author (Leydesdorf and Shin, 2010). While fractional counts may help to adjust for the productivity per researcher, it is obvious that simple meta data cannot precisely tell us the contribution of a single researcher to a scientific paper. As an example, in many scientific disciplines it is traditionally the "first author" who did most of the work, yet there is no satisfactory way to deal with this behavioral bias by using quantitative methods only. A further aspect concerns the incentives for researchers, which result from quantitative counting. While bibliometric studies may be relevant for individuals' tenure track and reputation, researchers could be

incentivized to increase their respective productive output by publishing their results in parts, also known as "salami-slicing" (Russell and Rousseau, 2002; Geuna and Martin, 2003). Sole publication counts may give a hint about the quantitative output of a researcher, but they fail to reveal the qualitative value. Besides qualitative reviews, journal rankings may help to partly overcome this problem. Since the incentive for publishing in a higher ranked journal with blind reviews is relatively high, one may assume that the selection process of top journals leads to higher quality publications with correspondingly higher scientific value. From a methodological perspective, this can be achieved by using the respective journal ranking. One approach to implement approach would be to simply the publications as proposed in equation 2.1.

$$Pu = \sum_{i=1}^{n} w_i p_i \qquad (2.1)$$

where the overall publication count Pu for publications in i journals is simply the sum of the authors' publication counts p_i in journal i times the rank weight w_i of journal i. The method becomes more ambiguous as soon as more journal rankings are available (Okubo, 1997). In this case, one can calculate aggregate weights or try to evaluate for each journal ranking separately. As regards impact measurement, the completeness of data plays an important role for sound productivity measurement. While the present data set focuses on German publication only, some German authors though they could have been highly productive in the area of environmental economics may be excluded if they have published abroad. In cases, I refer to more international scale studies of the field which rely mainly on ISI or Scopus data bases (e.g. Costanza et al., 2004; Ma and Stern, 2006).[5]

- Collaboration
  Knowledge generation often requires the cooperation and collaboration of several researchers in a joint project. Over all disciplines, bibliometricians observe increasing collaboration and cooperation across national and institutional boundaries (Subramanyam, 1983; Bordons and Gómez, 2000). Collaboration appears in bibliometric data through co-authorship which is a good, measurable proxy for collaboration. As Bordons and Gómez (2000) state, collaboration is particularly important in fields with high technological and management requirements. Hence, a high level of cooperation in the area

5 For a more detailed view on issues of productivity measurement, Van Raan (2005); Müller (2012) should be taken into consideration.

of environmental economics, as a transdisciplinary field of research, is expected. The national focus of this analysis will not allow for an international evaluation. Nevertheless, this study may contribute to the question of how national collaboration has evolved over time. Therefore, I follow the approach of Ajiferuke et al. (1988). The distribution of collaboration data is usually positively skewed. In general, one may find many publications with a few authors but only a few publications with many authors. It can be therefore difficult to show increases over time by simply using arithmetic means or medians (Havemann, 2009). Ajiferuke et al. (1988) suggest calculating a collaborative index (*CI*) as a robust indicator. Let n_j be the number of *j*-authored publications, N_p is the total number of publications, and *J* the greatest number of authors per document, then

$$CI = \frac{\sum_{j=1}^{J} j n_j}{N_p} \tag{2.2}$$

While the index is easy to calculate, it also has two major drawbacks. First, the index counts single author publications with a positive weight. Second, the index has no limits and may therefore be difficult to interpret. Alternatively, one can also calculate the collaboration coefficient (*CC*) which can be interpreted as degree of collaboration with values between 0 and 1. The disadvantage is that the coefficient only considers the proportion of collaboration, without regarding the distribution (Ajiferuke et al., 1988).

$$CC = 1 - \frac{n_j}{N_p} \tag{2.3}$$

Evaluative methods

In contrast to the above descriptive methods, evaluative methods aim at describing the collaboration and connection between documents or researchers. Basically, there are two main approaches for the identification of coherent research communities. The first one is the application of keyword similarity approaches (e.g. Spertus et al., 2005). Keyword similarity approaches and combined keyword and citation data approaches require specified keyword lists (Braam et al., 1991a, b). However, word lists pose the risk of incompleteness and thwart the idea of asking researchers for relevant literature. The second approach makes use of citation data. As a key assumption, Citations are assumed to be a good indicator of scientific communication between researchers (Garfield, 1979). In citation analysis, the extraction of reference lists and the subsequent statistical evaluation is a common technique for measuring scientific output and impact. However, citation data can also be used to characterize network relationships and to find similarities among documents. To identify similarities between documents, Kessler (1963) introduced the concept of bibliometric coupling, and Small (1973); Marshakova (1973) were the first to independently present the concept of co-citation. A bibliometric coupling occurs when several authors cite the same documents. In contrast, co-citations appear when two or more documents are cited often by a third (See Fig 2.1). While bibliometric coupling provides a retrospective view, co-citation analysis is slightly more forward-oriented (Garfield, 2001). Thus, co-citation is considered to be well-suited for mapping new, arising fields of research. The co-citation analysis presented in sections 2.1 and 2.2 can be divided into four steps:

1. Data definition
2. Data representation
3. Data evaluation
4. Data visualization

The first step is to define the object to be studied along with selecting the corresponding data. Depending on the object to be studied, two different approaches within co-citation analysis have been developed. Many researchers aim to characterize author relationships and their respective networks. Classic Author Co-citation Analysis (ACA), introduced by White and Griffith (1981), considers certain author groups. The main assumption is that authors do not change their specialties and represent a certain research area. However, ACA studies also pose the risk of being biased when selecting relevant researchers or ignoring citations of co-authors. It was mainly proposed for reasons of simplicity when working with Science Citation Index (SCI) data (White and Griffith, 1981). In contrast, prior Document Co-citation Analysis (DCA) maps all relevant documents and the citation relations among them. The resulting map shows the relationships of documents and topics rather than authors' collaborations (Schäffer et al., 2011). For this reason, the DCA approach is preferred.

The second step is to define an appropriate form of data representation. While all reference lists of documents can be easily represented by graphs, a more computational form is needed for data evaluation. Depending on the algorithms and data bases applied, there are many possibilities for representation. Schäffer et al. (2006), for instance, store the extracted data in Excel spreadsheets and use the pivot table function. For a simple application of well-known algorithms, an ordinary adjacency matrix A is applied, which defines citing documents i and cited documents j. If there is a tie between i and j the value is defined as one, otherwise the value is zero. A document cannot cite itself by definition. Thus, adjacency matrix A represents all collected reference lists.:

$$A \text{ with } a_{ii} = 0, \ \forall i > 0 \tag{2.4}$$

The third step includes data processing and evaluation. Co-citation is a common concept when measuring similar documents. Two or more documents are considered to be related to each other if they are cited within a third document. As shown in Figure 2.1 and mentioned above, co-citation can be easily represented in graph notation, showing two documents cited by a third author. For calculation, co-citation can be observed by drawing an arrow from the cited document backward to the citer and subsequently from the citer to another cited document. In Figure 2.1 this backward-forward operation derives two paths which are represented by dotted and dashed arrows (the path back to the object itself is not shown).

Figure 2.1.: Concept of bibliometric co-citation and coupling in graph notation.

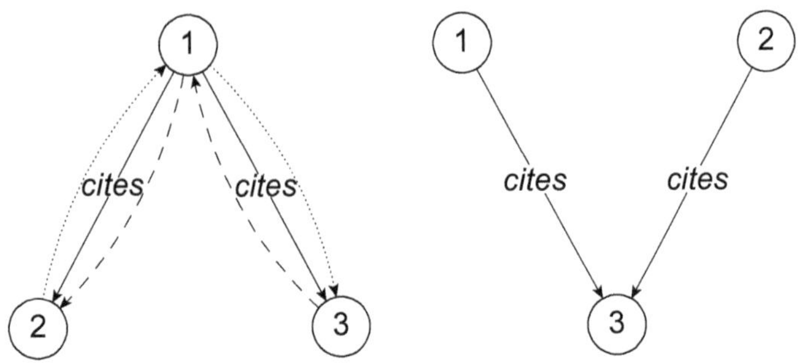

Given the adjacency matrix A, the operation can be calculated by multiplying A transposed with A, resulting in the co-citation matrix:

$$Coc = A'A \tag{2.5}$$

The co-citation matrix indicates all documents *i* and *j* that have been co-cited in absolute counts. If document *k* was cited, there is at least one path to the citer and one back to document *k* itself, hence the diagonal is $Coc_{kk} \geq 1$.

Based on the co-citation matrix, scientists have developed a large variety of methods for clustering groups of similar documents or authors. Early approaches directly applied absolute co-citation counts derived from the co-citation matrix (Small and Griffith, 1974). While this method is still appropriate for small numbers of observations, it does not take into account that the probability of co-citations depends on the number of co-citation possibilities. To overcome this problem, researchers extended the old approach by introducing relative similarity measures. Scientists frequently rely on the Jaccard-index, Salton's cosine measure or Pearson's correlation coefficient as relative similarity measures. The concept of the Jaccard-index is straightforward and can be directly expressed in set notation. It is defined by the intersection of two sets divided by their total (Small, 1973; Hamers et al., 1989).

$$J_{ij} = \frac{|\mathbb{C}_i \cap \mathbb{C}_j|}{|\mathbb{C}_i \cup \mathbb{C}_j|} = \frac{Coc(i,j)}{Cit(i) + Cit(j) - Coc(i,j)}, J_{ij} \in [0,1] \tag{2.6}$$

Salton's measure follows the idea of comparing two similar vectors. Vectors that have the same direction can be considered as equal, whereas dissimilar vectors are perpendicular. In matrix notation, Salton's measure is defined as:

$$S_{ij} = \frac{Coc(i,j)}{((cit(i) \cdot Cit(j))^{\frac{1}{2}}} = \cos\alpha, S_{ij} \in [0,1] \tag{2.7}$$

All these approaches lead to symmetric similarity matrices which define absolute identical objects as 1 and differing objects as 0. Defining thresholds is an efficient and simple way to extract clusters from similarity matrices. Moreover, similarity matrices are predestined for further multivariate analysis methods such as multi-dimensional scaling, hierarchical clustering techniques or factor analysis. This paper follows a similarity measure called CoCit-Score, which was first introduced by Gmür (2003).

$$CoCit_{ij} = \frac{Coc^2(i,j)}{min(Cit(i),Cit(j)) \cdot mean(Cit(i),Cit(j))}, CoCit_{ij} \in [0,1] \tag{2.8}$$

Compared to Jaccard's or Salton's measure, the goal of this measure is to raise robustness by decoupling the co-citation measure and the absolute citation counts *i* and *j*. It combines mean citation frequencies and minimum citation frequencies as relative terms (Gmür, 2003). The resulting CoCit-matrix does not fit the definition of a metric in a mathematical sense and is consequently referred to as a score. As long as no classic metric clustering methods (e.g. metric multidimensional scaling) are subsequently applied to the CoCit matrix, the score characteristic has no meaning and the score can directly be used in combination with determined thresholds. Gmür (2003); Schäffer et al. (2006) propose a threshold level of 0.2 as the minimum distance level for Cocit-Scores.

The last step in co-citation analysis is presenting and visualizing the evaluated data. The improvement of visualization algorithms is still an object of research (Moody et al., 2006). In this dissertation, I directly visualize the CoCit-Score similarity matrix. Hence, I use a "spring model" algorithm which minimizes the "spring" energy levels of two vertices (Eades, 1984; Fruchterman and Reingold, 1991). Cocit-matrix calculations were directly exported to Gephi visualization software (Bastian et al., 2009). In comparison to previous applied visualization methods, similarity relationships can currently be highlighted by edge thickness (Babl et al., 2014). All graphs are presented in undirected layout and do not follow arrangement patterns. While the algorithm by Fruchterman and Reingold (1991) delivers uniform edge lengths, Gephi allows to manually adjusting some vertices for legibleness.

2.1.2. Data

The data set of this dissertation is based on Babl et al. (2014) and extends the results for the three most recent years available (2010-2012). Therefore, the data set covers the time period between 1990 and 2012. In order to select relevant publications with scientific reputation, the journals included must have at least a VHB-Jourqual 2.1 ranking of "D" (Schrader and Hennig-Thurau, 2009). Through this constraint three journals in German speaking research communities are identified: (i) Zeitschrift für Umweltpolitik & Umweltrecht (ZfU), (ii) GAIA, and (iii) Zeitschrift für Energiewirtschaft (ZfE). ZfU, founded in 1978, enjoys a high level of acceptance and is the journal with the highest VHB-rating "B" in the data set. Established in 1992, GAIA, has a broad focus on ecological studies. The journal has a somewhat different structure than comparable journals. As a rule, there are more and shorter articles in GAIA per issue. However, many issues of the journal concentrate on topics which are highly relevant not only for ecological studies but also for environmental economics. Finally, Zeitschrift für Energiewirtschaft is a journal with an emphasis on the energy industry. The journal was added to the sample because of highly topical articles on renewable energies and related economic policies.

Before coming to the bibliometric results, this paragraph briefly summarizes descriptive statistics of the data set. The data set considers a total of 1320 documents which were cited between 1933 and 2012, comprising 714 citing documents and 910 cited documents. Table 2.1 gives a quick overview. Not surprisingly, the three abovementioned journals, ZfU, GAIA, and ZfE, account for the lion's share of the data set. However, there is a clear dominance of GAIA in the period from 2000 to 2012, which can partly be explained by the different structure of the journal. This disproportionate increase in the number of GAIA publications may lead to some bias which can be treated by the abovementioned methods (e.g. use of ranking weights). In addition to the documents of the three selected journals, the data set comprises several cited documents from VHB-relevant Journals. Interestingly, "Perspektiven für Wirtschaftspolitik", a journal focusing on economic policies, has taken on a higher level of importance in the field of environmental economics. Moreover, "Die Betriebswirtschaft", "Schmalenbach's Zeitschrift für betriebswirtschaftliche Forschung (zfbf)" and "Zeitschrift für Betriebswirtschaft" are German speaking journals with significant impact in areas of environmental economics. In contrast, other economic journals lost citation shares, which might be a sign of increasing professionalization and specialization of the delineated research field.

Table 2.1.: Publications included. The data set completely covers the three journals with a focus on environmental economics (ZfU, ZfE, GAIA) from 1990-2012. Additionally, the set includes all relevant articles published in VHB-ranked and German journals with a ranking of D and higher.

	Period	**1933-1999**	**Period**	**2000-2012**
Betriebswirtschaftliche Forschung und Praxis (BFUP)	16	(2.8%)	2	(0.3%)
Die Betriebswirtschaft (DBW)	28	(4.8%)	7	(0.9%)
Die Unternehmung	9	(1.6%)	1	(0.1%)
Finanzarchiv	6	(1%)	-	(-)
GAIA	163	(28.1%)	397	(53.65%)
Jahrbuch für Wirtschaftswissenschaften	4	(0.7%)	2	(0.3%)
Jahrbücher für Nationalökonomie und Statistik	8	(1.4%)	2	(0.3%)
Kredit und Kapital	2	(0.3%)	-	(-)
Marketing ZFP	5	(0.9%)	2	(0.3%)
OR Spectrum	1	(0.2%)	-	(-)
Perspektiven der Wirtschaftspolitik	-	(-)	10	(1.4%)
Schmalenbachs Zeitschrift für betriebswirtschaftliche Forschung (zfbf)	14	(2.4%)	8	(1.1%)
Steuer und Wirtschaft	5	(0.9%)	-	(-)
Wirtschaftsprüfung (Wpg)	5	(0.9%)	-	(-)

	Period	1933-1999	Period	2000-2012
Zeitschrift für Arbeits- und Organisationspsychologie	2	(0.4%)	-	(-)
Zeitschrift für Betriebswirtschaft (ZfB)	19	(3.3%)	9	(1.2%)
Zeitschrift für Controlling & Management	-	(-)	2	(0.3%)
Zeitschrift für Energiewirtschaft (ZfE)	109	(18.8%)	133	(17.9%)
Zeitschrift für Planung und Unternehmenssteuerung (Journal of Management Control)	1	(0.2%)	1	(0.1%)
Zeitschrift für Verkehrswirtschaft	1	(0.2%)	-	(-)
Zeitschrift für Wirtschafts- und Unternehmensethik	1	(0.2%)	2	(0.3%)
Zeitschrift für Umweltpolitik und Umweltrecht (ZfU)	178	(31.8%)	160	(21.6%)
Zeitschrift für öffentliche und gemeinwirtschaftliche Unternehmen (ZögU)	3	(0.5%)	2	(0.3%)
Total	580		740	

2.1.3. Results

As described in the methods section, the results are divided into two parts. The first discusses the main results extracted from descriptive methods, while the second focuses on the evaluative visualization of the scientific networks with the highest impact.

Results from descriptive methods

The descriptive analyses of the data set start with an overview of authors' productivity. The main results are listed in Table 2.2. The top 20 authors were selected and sorted by the first column "publications weighted score", which reflects the overall productivity and is calculated from equation 2.1. As stated above, the score is based on the fractional number of publication counts and the respective VHB-ranking. Furthermore, the overall amount of citations received from all documents within the given data set is calculated. Since citation data may still include self-citations, the overall citations corrected for self-citations are included as well. In the last column, the number of publications based on VHB-Jourqual ranks is shown. As can be seen easily, most ranks are congruent with fractional publication counts and the influence of the VHB-Jourqual weights is rather low. However, publication productivity does not always go hand in hand with citation counts. Especially, Jochen Jaeger and Martin Scheringer received much more citations from the scientific audience than colleagues. One possible explanation is that the authors published mostly in "GAIA" which has a strong journal self-citation behavior. The second possibility is that the mentioned authors wrote some introductory texts which are cited more frequently than regular articles. Overall, the citation counts seem to be rather (very) low in comparison with American studies or other fields of research. Thus, the first results also reflect the size of the research field, which can be characterized as rather small. Before proceeding to the results of the most frequently cited documents, I will provide a closer look at the work of the 3 top researchers. hence, their work is briefly summarized by an in-depth examination of the authors' articles and web presence. The first author, Erik Gawel, published several articles in highly rated journals and his interests cover a wide range of environmental and economical issues from environmental theory to (water) pollution. In his recent publications, he critically analyzed the latest developments of the "turnaround" in German energy policies. The second author, Armin Grunwald, focuses on technology assessment and system analysis. In accordance with his main research focus, Armin Grunwald published several articles on the environmental impact of several (environmental) technologies. There are, for instance, articles on technologies such as soil conservation, chemicals and their political regulation or Carbon Capture and Storage (CCS). An economist with a strong emphasis on environmental theory, Alfred Endres, who is the third author, wrote several highly ranked articles on the theory of sustainability, environmental risks, and problems of (monetary) evaluation of the environment. Furthermore, his research connects economic rational and necessities in environmental politics.

In addition to a view on authors' productivity, the descriptive analysis also includes the listing of the most cited documents. Table 2.3 lists the three most cited documents for every selected journal and the total ranking in descending order. The separation of journals helps account for the different citation behavior in the different journals. In ZfU, two documents which deal with policy instruments against atmospheric pollution are identified. The first was written by Tietenberg (1980), the first deals with the use of discharge permits as an instrument of air pollution regulation. The second by Hansjürgens and Fromm (1994) discusses success conditions for emission certificates using the example of the US-Clean Air Act of 1990. Especially in combination with the results in the evaluative part (see cluster 11 below), ZfU had a strong focus on air pollution mitigation and emission trading policies. The document listed last addresses success conditions in environmental politics, especially at the international policy-making level (Jänicke, 1990). It fits well to the aforementioned document, namely, on the efficiency of environmental instruments at the policy level (rather than politics and polity level). The second journal, ZfE, shows somewhat different results. Not surprisingly, the focus is more on energy policies and the energy industry. The most cited source in my data set is by Drillisch (1999) who deals with quota regulation policies for regeneratively produced power. While Germany implemented the Erneuerbare Energien Gesetz (EEG) on the basis of feed-in-tariffs, quota regulation would have been and still could be an alternative for policymakers. The main (economic) argument in favor of quota regulation is the advantage of increasing competition in contrast to rather inefficient fixed tariffs. Riechmann (1995) wrote a frequently cited article on price-cap regulation. The energy industry, typically a sector with extremely high fix costs and comparable low variable costs, easily leads to natural monopolies, thereby inducing higher prices with welfare losses. The topic is not originally an environmental issue, but it makes clear that the regulation of renewable energy markets may also be affected by the issue of natural monopolies and the regulation of network access tariffs. Similarly, the third document deals with (economic) benchmarks as a political instrument to evaluate the efficiency of electricity markets. Economic benchmarks may be an important tool to discipline market participants and to coordinate the price control mechanisms (Ajodhia et al., 2003). In GAIA, the most cited source

Figure 2.2.: Collaboration measures. The collaboration index (CI) is shown on the left, the collaboration coefficient (CC) on the right.

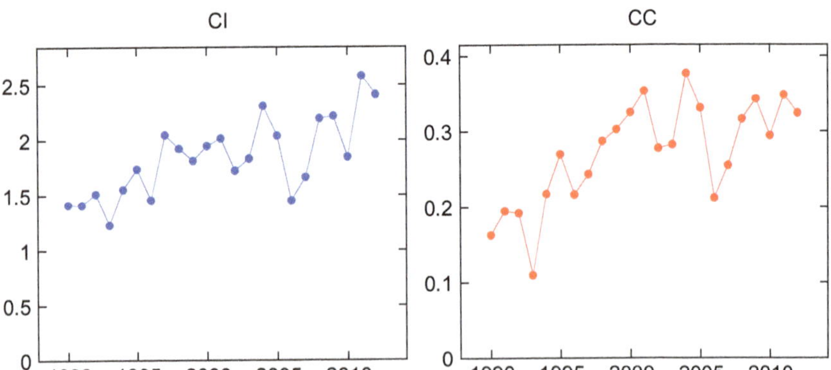

is by Jaeger and Holderegger (2005) who address the issue of landscape fragmentation. From an outsider's perspective, it is astonishing that landscape fragmentation could create more attention than many other, highly topical, issues in GAIA. The second and third frequently cited articles are Primas (1992) and Gethmann and Mittelstraß (1992). Both documents appeared in the first issue of GAIA, and are probably, but not necessarily, widely cited due to their exposed position.

For the evaluation of co-authorship collaboration I refer to Figure 2.2, which includes graphs for CI and CC. As can be seen easily, both indicators have been on the rise since 1990. CI and CC indicate that collaboration almost doubled. Especially CC, which can be interpreted as a proportion of collaborative work, shows collaboration levels have doubled from approximately 15% in 1990 to 30% in 2012. The research field can therefore be characterized as "normal" - the increase of scientific collaboration can be observed in almost every field of research and indicates a professionalization and an increase of complexity in the respective field.

Table 2.2.: Author productivity list. The table lists the most productive authors ranked by publications weighted score. In addition, the (fractional) number of publications and citations (without self-citations) are provided. The last column shows the VHB-relevant publications found in the data set.

Pos.	Author name	Publications weighted score	Fractional number of publications	Number of publications	Citation counts	w/o self-citation	Journals published
1	Erik Gawel	62.35	9.25	12	16	11	8xB, 1xC, 3xD
2	Armin Grunwald	42.16	7,6	9	15	14	1xB, 8xD
3	Alfred Endres	37.92	6,66	9	12	11	5xB, 1xC, 3xD
4	Ortwin Renn	37.24	7	10	18	14	10xD
5	Peter Kreuzberg	34.92	5	6	7	5	1xC, 5xD
6	Thomas Dyllick	34.91	5.64	8	10	7	2xB, 1xC, 5xD
7	Stefan Schaltegger	32.24	5,33	9	12	8	4xB, 5xD
8	Ulrich Müller-Herold	31.81	5.98	8	9	8	7xD
9	Heinz Eckart Klingelhöfer	31.1	5	5	3	0	2xB, 3xD
10	Martin Scheringer	29.89	5.62	11	38	26	11xD
11	Michel Roux	29.42	5.53	10	10	6	6xD
12	Carl Christian von Weizsäcker	28.24	5	5	6	6	5xD
13	Jochen Jaeger	27.93	5.25	11	42	36	11xD
14	Friedel Bolle	26.64	4.5	5	4	2	1xB, 4xD
15	Rolf Peter Sieferle	26.6	5	5	9	9	5xD
16	Dieter Steiner	26.6	5	5	4	3	5xD
17	Reimund Schwarze	25.61	3.33	6	8	8	4xB, 3xC
18	Manuel Frondel	20.4	3.33	6	5	2	2xB, 4xD
19	Ingrid Kissling-Näf	19.68	3.7	5	4	4	5xD
20	Irmi Seidl	18.78	3.36	8	11	9	1xB, 7xD

Table 2.3.: Most cited documents. The list separates the results for each delineated journal. Ranking according to absolute citation counts w/o self-citations.

Pos.	w/o self-citation	Absolute citation counts	Document
Zeitschrift für Umweltpolitik und Umweltrecht (ZfU)			
1	5	5	Tietenberg, T. H. (1980): Transferable Discharge Permits and the Control of Air Pollution: A Survey and Synthesis. In: ZfU. 1(1980), pp. 477-508.
2	4	5	Hansjürgens, B.; Fromm, O. (1994): Erfolgsbedingungen von Zertifikatelösungen in der Umweltpolitik am Beispiel der Novelle des US-Clean Air Act von 1990. In: ZfU. 4(1994), pp. 473-505.
3	4	5	Jänicke, M. (1990): Erfolgsbedingungen von Umweltpolitik im internationalen Vergleich. In: ZfU. 13(1990), pp. 213-232.
Zeitschrift für Energiewirtschaft (ZfE)			
1	7	7	Drillisch, J. (1999): Quotenregelung für regenerative Stromerzeugung. In: Zeitschrift für Energiewirtschaft. (1999), pp. 251-274.
2	5	7	Riechmann, C. (1995): Price-Cap Regulierung. In: ZfE. (1995)2, pp. 157-167.
3	5	6	Ajodhia, V., et al. (2003): Benchmarking and its Applications. In: ZfE. 27(2003)4, pp. 261-274.
GAIA			
1	7	8	Jaeger, J.; Holderegger, R. (2005): Schwellenwerte der Landschaftszerschneidung. In: GAIA. 14(2005)2, pp. 113-118.
2	7	7	Primas, H. (1992): Umdenken in der Naturwissenschaft. In: GAIA. 1(1992)1, pp. 5-15.
3	6	6	Gethmann, C. F.; Mittelstraß, J. (1992): Maße für die Umwelt. In: GAIA. 1(1992)1, pp. 16-25.
Overall			
1	7	8	Jaeger, J.; Holderegger, R. (2005): Schwellenwerte der Landschaftszerschneidung. In: GAIA. 14(2005)2, pp. 113-118.
2	7	7	Drillisch, J. (1999): Quotenregelung für regenerative Stromerzeugung. In: Zeitschrift für Energiewirtschaft. (1999), pp. 251-274.
3	7	7	Primas, H. (1992): Umdenken in der Naturwissenschaft. In: GAIA. 1(1992)1, pp. 5-15.

Results from evaluative methods

In this section, interconnections and relationships between documents with the highest impact scores in terms of co-citation counts are analyzed. Figure 2.3 directly shows the results. Out of the initial 1320 documents in the data set, 75 documents in 16 clusters have at least been co-cited twice. The CoCit-Scores reveal the strength of relationships between the 75 documents and are highlighted by edge thickness. Most clusters tend to be rather small with a median of 4. The largest cluster has a size of 10. However, one should be aware that collaboration networks may be much more extensive and appear to be strongly filtered by the parameter settings. In order to analyze the content and subjects of the clusters, I identify the most central document in each cluster first. While this task may be more complex to calculate for bigger clusters, it is much easier for the clusters in Figure 2.3. Nonetheless, in literature some centrality measures are known, eigencentrality or degree centrality for instance, but for the purpose, the count of degrees, that is the count relationships or edges in relation to all cluster edges, seems to be sufficient. However, it is not possible to determine the "most" central document. Once the most central source has been detected, the next step is to evaluate the respective headlines, abstracts, and sometimes even the articles themselves in order to interpret what the clusters are concerned with. Table 2.4 corresponds to Figure 2.3 and summarizes the obtained results.

The following section provides brief overview of the development of clusters over time. Therefore, the citation year rather than the publication year is used in order to reveal when clusters were current. The oldest cluster, cluster 11, which was cited from 1991 to 1995 and deals with environmental licenses and emissions trading. The most central source is Huckestein (1993), which deals environmental licenses and the political quantity control. However, the oldest cited source of the cluster is Tietenberg (1980) and illustrates the long lead time before the instrument was actually established. The second cluster in the timeline (1995-1996) is cluster 2 which deals with conservation of habitats and biodiversity. While one can still observe a lot of papers on that issue, the five documents mentioned here had the highest impact in the field. Baur (1995) is the most central source and discusses threats to flora and fauna caused by humans. At the same time, from 1995 to 2001, a cluster appears that regards price cap regulations of natural monopolies (cluster 7). Especially energy markets tend to be natural monopolies and price cap regulation was discussed as a political instrument in order to reduce possible welfare losses. In the context of environmental economics this applies to all (renewable) energy structures with high fix costs and low variable costs. The discussion was led by Zeitschrift für Energiewirtschaft and the most influential source in terms of degrees was Riechmann (1995). In cluster 6, I identify three documents that deal with the theory of sustainability. The concept of sustainability was originally derived from forest management considerations. However, studies in the 1990s analyzed how the concept of sustainability would work in an ever-growing economy and how sustainability impacts national accounts. The central source in this regard is Cansier (1995). Shortly after the government change to "Red-Green" coalition in 1998, cluster 3 on quota trading in the renewable energy sector appeared. The cluster is relatively small, Drillisch (1999) obviously started the debate. Cited mainly from 1999 to 2008, the cluster addresses how renewable energy quotas may help to increase the energy share in the total energy mix and how this will influence electricity prices. Cluster 14 includes eight documents and discusses general concerns of why environmental research does not have a higher level of relevance or appropriate significance. The discussion was initiated by the introductory text of Jaeger and Scheringer (2006). Cluster 15 was cited from 2005 to 2012 and deals with natural disasters, floods and flood control. It begins with articles by Schanze (2002) and Dombrowsky (2002), which directly handle the catastrophic event of the Elbe river flood in 2002. The cluster is a direct consequence of the climate change debate which is discussed later in cluster 5. Egner (2007) directly questions how climate change came into the debate. However, it is a surprising result, that German literature on environmental economics scarcely discussed issues of climate change so far. Taking into account its vast predicted economic effects, it was expected that policies of mitigation or adaption should have been discussed more deeply and profoundly. Then, three clusters which have been frequently cited in their given year are observed. The documents appear through strong cross-citational behavior in issues with featured main topics in GAIA. The first, cluster 16, appears in 2005 and discusses the problems of landscape fragmentation and urban sprawl. The second, cluster 10, was mainly cited in 2009 and deals explicitly with the Carbon Capture and Storage technology. The third, cluster 4, arose in 2012 and regards problems of pesticides and chemicals use. During the years 2007–2010 several studies were cited which dealt with the transdisciplinary nature of environmental economics (cluster 13). While the research field is tangent to many other scientific disciplines, such as economics, law, politics, or engineering, it demands much more integrated knowledge management. Transdisciplinarity may also be part of the answer as to why environmental research does not contribute more to environmental issues. Interestingly, Cluster 9 was cited from 2009 to 2010 and sceptically discusses the use of scientometric and bibliometric methods for the evaluation of scientific output. Nevertheless, there is no direct link between environmental studies and the content of the cluster but it shows that all scientific areas become more and more affected by scientometric methods and the wish to quantify the output of researchers. This also applies to environmental researchers who cannot exclude themselves from a critical discussion of the negative incentive effects. Cluster 8, cited from 2010–2012 reactivates a very old topic of environmental economics: The difficult valuation of environmental goods and the value of a healthy environment. However, the cluster does not link to older documents of the field and creates a new cluster. Cluster 1 opens a new discussion on sustainability and inquires whether politics or consumer behavior may lead to the realization of a sustainable economy. The last cluster, cluster 12, regards the use of virtual water and water footprints as an instrument to increase efficiency of water use. The core idea is to sum up the consumption that is needed for products and services and to generate water trade balances. As an example, nations may be able to reduce water consumption by importing food. With the help of the so called virtual water trade it may be possible to politically address water problems in regions with water shortages. However, the concept has also several drawbacks and pitfalls which are mainly discussed in Biewald (2011).

Table 2.4.: Table of cluster. The table can be assigned to the network in Figure 2.3. The cluster number directly refers to numbering of the network. The cluster size describes the amount of documents of the respective cluster. The citation period covers the years when the cluster was cited. Column most central document refers to the most central article in terms of degrees. The last column contains content information.

Cluster number	Cluster size	Citation period	Most central document (Degree centrality)	Main topics/Keywords
1	8	2011-2012	61, 62, 63 (0.86)	Sustainability: Politics and consumer's responsibility
2	5	1995-1996	46 (1)	Biodiversity
3	4	1999-2008	2 (1)	Quota trading in the renewables industry
4	4	2012	74 (1)	Ecotoxicology and Pesticides
5	5	2008-2009	3 (1)	Climate change
6	3	1996-2000	5 (1)	Theory of sustainability
7	3	1995-2001	51 (1)	Price-cap regulations for natural monopolies
8	4	2010-2012	- (all 1)	Valuation of environmental goods/ nature
9	3	2009-2010	- (all 1)	Culture of Knowledge generation and citation indices

10	4	2009	36 (1)	Carbon Capture and Storage (CCS)
11	3	1991-1996	54 (1)	Environmental licenses and emissions trading
12	3	2011-2012	68 (1)	Virtual Water and Water Footprints
13	4	2007-2010	41, 43 (0.67)	Environmental studies as a transdisciplinary field of research
14	8	2001-2008	27 (0.86)	Environmental research and solutions
15	4	2005-2012	21, 20 (0.67)	Natural hazards, floods and flood control
16	10	2005	12 (1)	Urban sprawl and landscape fragmentation

2.1.4. Discussion of Section 2.1

Section 2.1 presents a bibliometric analysis on German environmental economics based on 1320 articles and cited articles with a sample period from 1990 to 2012. The goal was to reveal the most productive authors, the most influential documents, and the development of collaboration and cocitation networks with high impact. The 20 most productive authors in terms of articles published in the respective journals are of importance for a first impression of the opinion leaders of the segment. However, comparable Anglo-American studies suggest that the impact of German-speaking communities on international communities is rather low (Ma and Stern, 2006). From a methodological point of view, it is important to bear in mind that ranking does not necessarily reflect the complete output of an author. Especially authors with a more international publishing focus may be subject to underestimating as only publications from German speaking communities are counted. For analyzing the most important documents, citation counts received are counted, while self-citations are excluded. The results are presented separately for each selected journal and reveal the different research foci. Highly cited documents in ZfU mostly deal with classic environmental (resource) economics. They focused on the field of air pollution reduction, emission and certificate trading and the determination of success conditions for environmental policies. However, when it comes to questions of renewable energy and natural monopoly policies of the energy sector, most cited articles can be found in Zeitschrift für Energiewirtschaft. The result is not really surprising, as the journal's focus lies on the energy industry. Nonetheless, it is interesting to note that both environmental and ecological journals, ZfU and GAIA, were not able to make the subject of "Renewables" their own. "GAIA", the third originally selected journal, has its focus on ecological issues and often touches themes of ecological economics. In many ways, the individual articles reflect the uncertain and transdisciplinary character in ecological economics (also compare to Castro e Silva and Teixeira (2011)). Highly cited documents reflect this complexity or cope with concrete ecological issues, such as landscape fragmentation. Generally, higher overall citation counts are observed in GAIA than in ZfU and ZfE which may be partly explained by indexation in the science citation index, along with easier electronic network access. Furthermore, an important, highly cited source outside the three selected journals cannot be identified, a fact that may be an indication of completeness of the data set. "Blind spots" in the data set, which may occur due to unsuitable journal selection criteria can therefore be discounted to a large extent. The evaluation of collaboration reveals that environmental economics is a "normal science" with increasing coauthorship. There may be many reasons for this, but as the intensity and complexity of a research field increases, it is only natural to assume that collaboration will increase accordingly. This finding is also underlined by the fact that 43% of the documents are included until 1999 but 57% after 2000, indicating an increase in the output.

The evaluative part of this study identifies 76 highly co-cited documents in 16 clusters. The analysis also reflects the development of the field over time. In order to determine and interpret the content of the clusters I focused on the abstracts of the most central sources. From an external view, it is astonishing that long-term research networks on climate change or the valuation of externalities cannot be detected. Instead, one can find a series of articles that attract attention for shorter time periods and appear to be processed in bundles. It may also be an interesting observation that the network of co-cited documents is rather small in comparison to other fields of business or economic research. The reasons are open to speculation, but it is obvious that some highly relevant documents may not be available electronically and may hence be cited less frequently than comparable Anglo-American publications. Furthermore, many publications are solely written in German and may not attract a broader international audience. The overall higher citation counts of GAIA underline this impression. The results are also surprising in another way: The German "Energiewende" and the transformation to clean technologies, such as e-mobility, causes huge transformation costs, which are not really addressed in the highlighted research clusters.

Overall, one can see a rough division of German environmental economics research in three parts which correspond to the selected journals. This division also reflects the differences between environmental and ecological economics (ZfU vs GAIA) and research that deals with (renewable) energy policies (ZfE). While all three address economic costs and risks of environmental destruction in different facets, it seems to be desirable to coordinate the results better and to identify the most important preliminary work. This section may help researchers to maintain the necessary overview.

In the next section, the perspective will shift from a general impact view of environmental economics to a more concise view on Cleantech and the detection of all relevant publications.

2.2. Detecting relevant literature for Cleantech applications

For the last two decades, the Cleantech sector in Germany has emerged as an important economic sector with remarkable progress and growth and it may be assumed that similar developments apply for the E-mobility industry in the near future. Limited natural resources, climate change and declining biodiversity have forced politicians, engineers and decision-makers to support new concepts of sustainable living and to develop alternative technologies. Governmental subsidies and stimulus packages have helped to advance new clean businesses, which has led to the formation of a new industrial sector in Germany. This process has brought about a high level of recognition for green policies and technologies. A wide variety of journals and newspapers as well as internet articles deal with new technology applications and aim to raise public awareness. By emphasizing Germany's world market leader position, German policy-makers try to benefit from the generated publicity.

However, the development of the sector has not been without difficulties. Over the past years, many of them came to face problematic and distressed situations which were partly caused by self-induced mismanagement but also but relying too much on governmental assistance. Solar modul producers, for instance, have been lagging behind in developments and were strategically not prepared for the strong Chinese competition and capacity building. An interpretation of this general impression could be that most businesses in an early stage greatly benefitted from engineering inventions and innovations in the past but experience administrative or regulatory macroeconomic deficits at later stages. The deficits observed raise the question of whether economists have addressed industry-specific problems early enough and offered sensible solutions to managers and policy-makers. To address this aspect and to obtain a complete overview of the delineated research field, I conduct in this section an additional analysis on the German Cleantech industry.

With the increasing importance of the Cleantech sector, one might assume that its scientific importance in (environmental) economic and business journals is increasing as well. However, the results of Section 2.1 do not indicate an increase of importance in German-speaking countries. Besides the CCS cluster, no high-impact cluster with explicit Cleantech content could be identified. The same applies to literature reviews of other authors. One reason for this could be the nature of the research field, which is in fact transdisciplinary. Environmental economics integrate various research fields such as economics, business administration, engineering, biology, and geology. Bibliometric studies therefore pose the problems of survey definition and journal selection. Another explanation can be found in the structure of leading citation data-bases. Most researchers rely on Thomson's SCI or Scopus when conducting bibliometric studies. Currently, neither data-base includes a sufficient coverage of the research from German-speaking countries.

In this section it is the goal to analyze the aforementioned data set more in detail. Methodology is slightly adopted to reveal all Cleantech relevant publications or in other words co-citation methodology is this time applied as a detection method. Therefore section 2.2 mainly covers two explicit research questions:

1. How can the most relevant literature for Cleantech industry be extracted and summarized?
2. How did literature on Cleantech change over time?

In order to answer the questions, I perform an additional co-citation analysis, but this time the method is applied as a sensitive detection method. The next section will shortly clarify the differences to section 2.1.

2.2.1. Necessary adaption of the methodology

In contrast to Section 2.1, it is this time the foal to detect all literature that deals with Cleantech applications. Depending on the parameters of the co-citation analysis, the method can sensitively extract complete community structure maps

Table 2.5.: Sensitivity parameters. The choice of the cocitation mapping method and parameter settings heavily depend on the particular application. The settings may vary from very sensitive topic detection to the detection of the most influential publications of a research field.

Parameter	Rather search sensitive	Rather impact oriented
Mapped entity	documents (DCA approach)	documents or authors (DCA or ACA approach)
Number of cocitations	1	≥ 2
Smallest cluster size	small (≤ 3)	large (≥ 3)
Threshold levels for similarity measures (Cocit-Score, Jaccard-Index, etc.)	lower threshold levels (no loss of information)	higher threshold levels (more relevancy)

or detect the most influential articles only. Hence, an adaption of the co-citation methods or of the parameters of the co-citation analysis is required in order to reveal the most influential sources. An overview to sensible research parameters is given in Table 2.5 and explained in more detail below.

As described above, one may perform a DCA or an ACA. As DCA describes every single document in contrast to the ACA, it can be seen as a superset. It is somewhat more accurate and particularly suitable for detecting single documents. The second and probably the most important parameter for eliminating less relevant documents is the number of co-citations needed in order to be included in the analysis. With increasing co-citiation numbers, the method becomes less and less search sensitive and eventually extracts the most coherent and interrelated documents, as seen in section 2.1. Here, *Coc* = 1 is determined, or in other words, every co-cited document becomes relevant for evaluation. Furthermore, the minimum allowed cluster size may play a role, as bigger clusters may indicate more relevancy but concurrently reduce sensitiveness. However, I still apply a cluster size as it clearly shows the context of single documents. The last point in Table 2.1 is the applied threshold level. Increasing threshold levels may help to identify more relevant clusters but do at the same time split off documents and may therefore mean a loss of information. Here, relevant thresholds can be seen by edge thickness and readers can easily see the importance of each connection.

2.2.2. Results

The results are presented in two subsections. The first subsection summarizes all results of co-citation analysis. It features a complete overview of environmental economics in German-language literature. The analysis also extracts all Cleantech relevant clusters. The second section draws a timeline of the delineated research field.

Co-citation analysis, mapping and visualization of research fields

The classification of research topics is one of the main advantages in co-citation analysis. The map of the delineated research field can be easily derived by the co-citation matrix. For mapping and visualization purposes, I apply Cocit-Score matrix at threshold 0.2. Fig. 2.3 illustrates the mapping results for my entire data set. The map consists of 637 co-cited documents, forming 77 co-citation clusters with an average size of 8.2. Each co-cited document is visualized by a graph vertex and is assigned to a unique numbered label.[6] Co-citation relationships are represented by edges between the vertices. Apparently, the number of clusters seems to be relatively high vis-à-vis comparable mapping studies. This can be partly explained by the low number of co-citations which is ≥ 1 in comparison with Figure 2.3 which has a co-citation number of ≥ 2. Another explanation could be that the threshold of 0.2 cuts edges between documents and leads to more fragmentation. However, even when clustering the co-citation matrix directly without any threshold, 59 clusters would remain. This can be interpreted as a good indicator for threshold robustness of the results. Most clusters are well localized and can be easily assigned to their corresponding research topics. In this case, the focus is on clusters dealing with clean technologies.

In contrast to my previous results (Babl et al., 2014), I identify two clusters more and one cluster less (excluded now by threshold), summing up for a total of 10 clean tech related clusters.[7] All clusters are listed in Table 2.6 and can be spotted on the map in Figure 2.4.

Cluster 1, consists of nine documents, mainly published in ZfU (8 out of 9). The main topic of the cluster is the impact of environmental innovations on economics. The cluster does not focus on concrete technologies but contains examples of sustainable technologies within the chemical industry or energy sector. Cluster 2 deals mainly with air pollution issues and emission certificates. It emerged in 1981 in the journal ZfU while discussing Americans' pollution rights. The later documents of this cluster discuss certificate implementations in European countries as well as certificate solutions in production planning. Since certificate solutions aim to internalize external costs of emissions, they may be considered as a key driver for the development of clean technologies. Cluster 3 can be assigned to the renewable energy sector in Germany. The cluster emerged in 1998 and started discussions on the pricing of green electricity (Weller, 1998). Furthermore, the cluster documents provide overviews on stimulation instruments and the offerings of green electricity for the green electricity market, which was still relatively young (Dreher et al., 2000; Bräuer, 2002). Cluster no. 4 is a tiny one, dealing with green energy stimulation and customers' preferences (Menges and Traub, 2008). In 2008, a new field of research arose and led to Cluster 5. Luhmann (2008) was the first to make CCS a subject of discussion in ZfU. In 2009, the journal GAIA had a special issue on CCS and contributed with several articles on that issue.

---

6 For reasons of space, a full list of all 637 documents is going to be available at http://tuprints.ulb.tu-darmstadt.de/

7 The excluded cluster dealt with coastal change by offshore wind plants.

Effects of environmental policies on innovation are discussed in Cluster 6. This cluster is a small one with a size of 4. The main topic is the influence of environmental policies on innovation (Hemmelskamp, 1999). Assessed from an external point of view, cluster no. 1 and 6 seem to be very similar in structure and research topics. Both deal with the complex "environmental policies" and impact on innovation and economics. In Cluster 7, issues relating to green electricity marketing instruments are discussed. Apparently, the relatively young market needed some adjustments on marketing instruments for attracting new customer groups (Graehl et al., 2001; Homburg et al., 1999; Zinnbauer, 2001). Cluster 8 seems to be very similar and deals with the acceptance of green electricity pricing (Gerpott, 2009; Rommel and Meyerhoff, 2009). In contrast to our previous findings, I spot two new Cleantech relevant clusters which emerged between 2011 and 2012. Cluster concretely discusses cost effects of the Renewable Energy Act. Amid growing concerns with the cost situation, the cluster directly discusses costs and possible needs for reform of the German "Energiewende". Very similar, the last cluster, No, 10, starts with the disastrous developments in the solar industry and questions whether the Energy transition went wrong.

While most clusters can be easily extracted and assigned to certain research fields, there are still clusters that did not dissolve well. Apparently, one cluster in the middle of the map is a conglomeration of several documents and should be discussed briefly. The cluster consists of 154 documents partly relevant to Cleantech applications (see Figure 2.5). For further investigation, it is possible to raise thresholds to split the cluster into several parts or to apply other algorithms as the Newman algorithm (Babl et al., 2014). However, by the examination of the documents' relations and their neighborhood within the remaining main cluster, one can identify two additional clusters related to clean technologies.

The first one comprises document nos. 195, 198, 199 and deals with renewable electricity and heat production, and is marked with an a) in Figure 2.5. The second emerged during the early 1990s and includes only two documents (no. 446, 449) which is below the demanded cluster size 3. Thus, it will not be counted as its own field of research. Nonetheless, the cluster is worth being mentioned because it deals with the control of industrial-scale facilities and environmental issues. The controversial case of the magnetic suspension railway "Transrapid" is a good example of a clean technology that reached large-scale dimensions.

To summarize the results from co-citation analysis, all Cleantech relevant clusters are summarized in table 2.6. Overall, I identified 87 documents within eleven clean technology related clusters. As expected, the largest proportion of documents and clusters deals with renewable energies. Nine out of twelve clusters are directly influenced by renewable energies. Two clusters show a rather abstract view of environmental policies and environmental problems on innovation (Clusters 1 and 6). Surprisingly, researchers were inspired by the CCS technology. Of all things, CCS technologies could become established within research literature while other concrete technologies, even those with higher application rates, could not. An explanation could relate to the nature of CCS which directly addresses problems of climate change by storing huge amounts of carbon.

Figure 2.3.: Visualization of high impact environmental economics cocitation clusters with Coc ≥ 2. Identification of 75 documents in 16 clusters. Cluster numbers help to assign the network to Table 2.4.

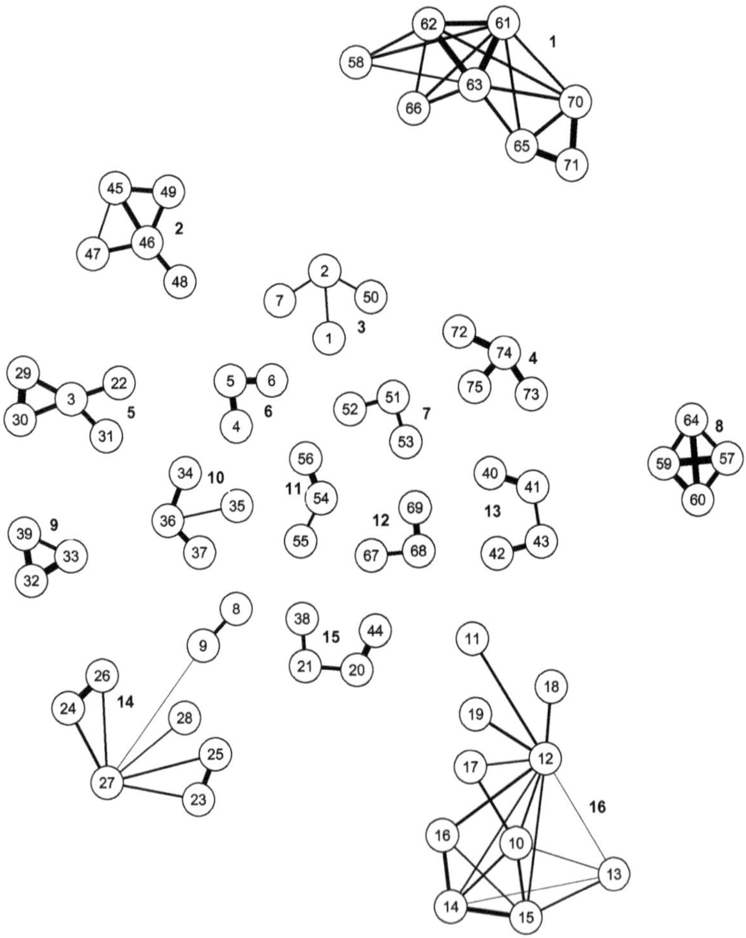

Figure 2.4.: Visualization of environmental economics co-citation clusters with Coc = 1. Search sensitive identification of 637 documents in 77 clusters. The ten Cleantech related clusters are highlighted in numbers from 1-10 and correspond to Table 2.6

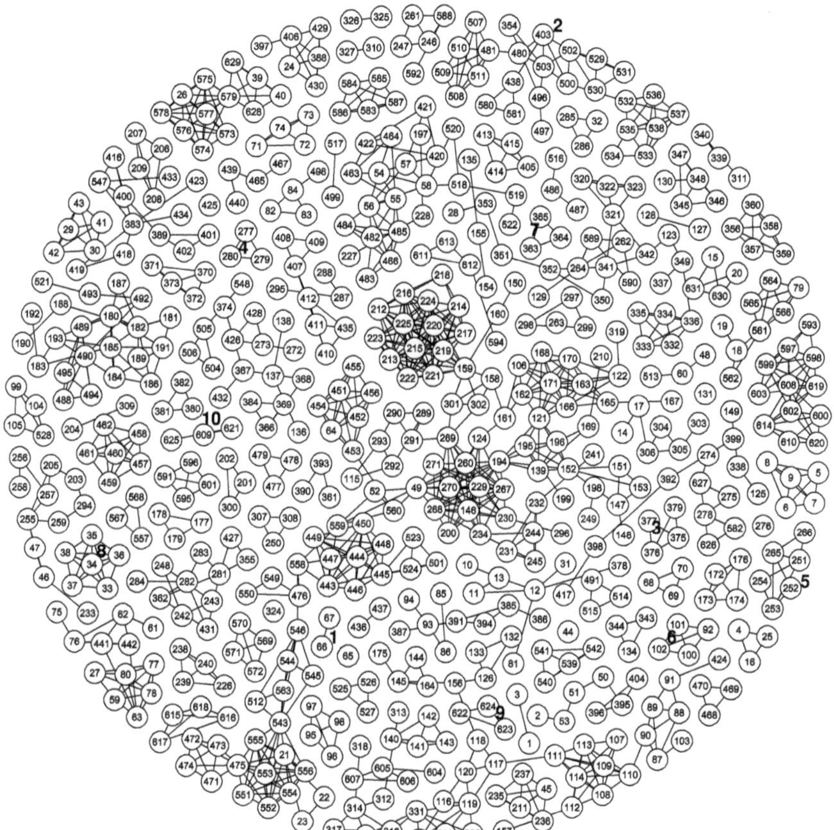

Figure 2.5.: Close-up of the large-sized cluster with several research foci.

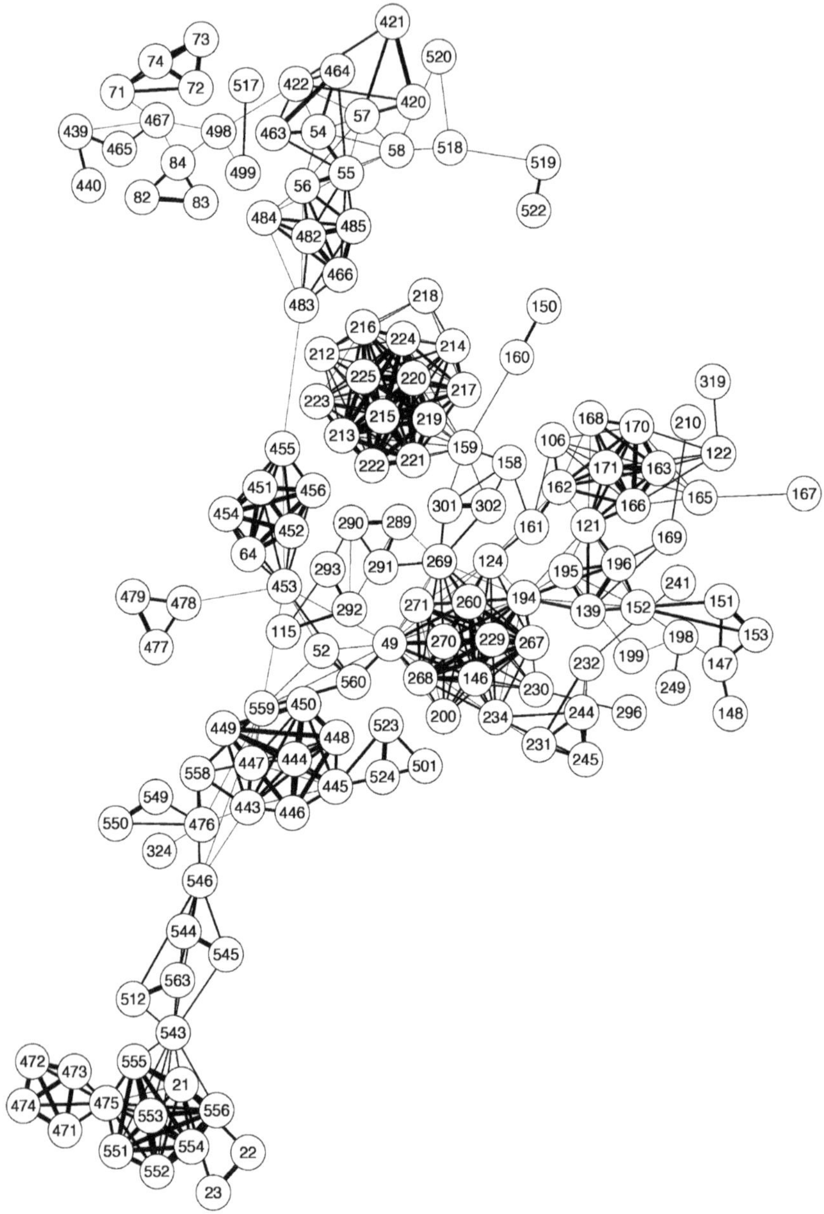

Table 2.6.: Results: Cleantech relevant clusters.

Cluster number	Cluster size	Period	Main topics/keywords
1	9	1987-2001	Innovation, technology, sustainability, conflicts
2	27	1981-2003	Air pollution, environmental licenses and certificates, ecological rationality, production and environmental issues
3	11	1998-2003	Renewables, green electricity
4	3	2008	Energy markets, requirement for green energy
5	6	2008-2009	Carbon Capture and Storage (CCS)
6	4	1999-2002	Influence of environmental problems on innovation
7	3	1999-2001	Electricity sales/Green electricity
8	6	1996-2009	Marketing and pricing/Green electricity
9	3	2011-2012	Costs of energy transition and climate change
10	3	2011-2012	Solar crisis and critical consideration of the Renewable Energy Act
(11 a))	3	2003-2005	Electricity and heat production/Renewables

Chronological survey

To address my last research question, I develop a chronological survey of Cleantech research for the last 20 years. Basically, in citation analysis there are two possible ways to analyze chronological changes in time. Given adjacency matrix A, one might reduce matrix A to all citations i within a given time span. Another approach would be an analysis of all cited documents j. While these approaches allow one to analyze citation behavior of researchers over time they do not contribute to my initial research goal which is a chronological mapping of the research field. To overcome this problem, I use the publication dates of all documents in my data set. Table 3 already contains all time information of Cleantech related clusters. For completeness, I extend the considered clusters with relevant single documents for gaining a better picture of the field.

After realizing that Chlorofluorocarbon (CFC) destroy the ozone layer, the German industry played a significant role world-wide in the dispensing with CFC. Hofstetter (1992) explains the need for CFC-free refrigerators, which was in fact a very early Cleantech invention for consumer goods. At the same time, Wagner and Matten (1992) show environmental concerns with the magnetic suspension railway "Transrapid." Three years later, Pütz and Schmied (1995) expand the topic by showing political influence on industrial-scale systems. About ten years after the reactor catastrophe in Chernobyl, it seems that people lacked confidence in industrial scale solutions for environmental problems on earth. At the same time, economists seriously questioned the need for economic growth and began to develop the idea of sustainable economies (compare with Cluster 1). Binswanger (1994) directly addressed the energy intensity of industrial structured economies and raises the question of energy efficiency. One year later he described possibilities for sustainable development (Binswanger, 1995). Sachse and Bach (1994) already discussed potentials and costs of solar energy. Radke (1996) is the first who discusses the economic aspects of "sustainable technologies."

Since 1998, there have been many articles on renewable energy. Changes in legislation with the renewable energy act in 2000 likely helped to boost this new industrial sector. The new product "green energy" obviously triggered marketing issues in terms of pricing and customer acquisition. Green electricity issues are still up-to-date (see Cluster 4). During the last decade, some more concrete articles on specific technologies have appeared.

Interestingly, there is almost no publication on solar energy during the first decade of the 2000s. Only Gisler and Pohl (2000) report on the Zurich solar power exchange.

This view changes when it comes to wind power. Gesang and Wiesenthal (2003) are the first to ask whether the expansion of the wind energy sector makes sense. The rentability of offshore wind parks in the German North and Baltic Seas was analyzed by Jeske and von Hirschhausen (2005), and the risks involved by Madlener et al. (2009). Meisel et al. (2009) describe the development of onshore wind parks. However, with the increasing use of wind power, new problems arose. Kannen and Burkhard (2009) examine the problem of landscape changes in coastal areas. Ohl and Eichhorn (2008) focus on problems of scarce land areas and discuss sustainability issues caused by the re-powering of wind turbines.

Within my data set, there is only one source concerned with biogas (Schidler and Sotoudeh, 2005). The first paper on geothermal use appeared in 2009 (Wenzel et al., 2009).

Outside the renewable energy sector, there is only one highly topical issue, namely sustainability in the automotive industry. Canzler et al. (1993); Knie, A. and Berthold, O. (1996) already questioned the use of automobiles for efficiency reasons back in the 1990's and discussed the application of E-mobility. In 2005, Henseling (2005) focused on the sustainability and mobility issue. More recent documents deal with the energy conservation issue in mobility. Frondel and

Peters (2007) question whether bio fuels could contribute to climate protection goals, and Reller et al. (2009) illustrate the risks of electric batteries using the example of cell phone batteries. The data enhancement for the years 2010-2012 reveals two new clusters which critically consider the effects of the renewable energy act. Frondel et al. (2011); Gawel and Korte (2012); Ziehm (2012) discuss issues of social and distributive justice of the Renewable Energy Act. The second cluster starts with the failure of the German solar industry and fuels a new critical discussion on the Renewable Energy Act (Eisenbeiß and Goetzberger, 2011).

2.2.3. Discussion of Section 2.2

In section 2.2 I examined the scientific output of environmental economists and business researchers on Cleantech over the past 20 years. As a main result of these examinations, one may conclude that business researchers indeed supplied a rather small and fragmented set of business knowledge for the Cleantech industry. The results underline that there is a gap between the valuation of clean technologies in public perception and their development as reported in economic literature. This peculiar result does not fit with Germany's claimed market position and with the fact that the technologies could contribute to the long awaited decoupling of economic growth and energy dependence. Moreover, the absence of knowledge could have intensified observed cases of mismanagement in the Cleantech industry.

For discussion, I would like to address my initial research questions:

- How can the most relevant literature for Cleantech industry be extracted and summarized?
- How did literature on Cleantech change over time?

The first question is aimed at an appropriate data collection and the methods applied. The collection of all relevant documents without available data-bases that cover the research field is a difficult task, especially when the research field cannot be easily defined by apt keywords. While many ACA studies rely on the extraction of author-lists from databases, there is no such developed data base in such a new and emerging research field. To overcome this problem, I completely mapped all relevant German literature on environmental economics. Unfortunately, this procedure causes additional efforts which go beyond the initial objective and collect superfluous data sets as well. On the other hand, keyword methods run the danger of being incomplete and missing relevant aspects. The advantage, however, is that the completeness of data allows external experts to select relevant data clusters themselves.

Co-citation analysis is a tested and approved classification technique for similar documents. Nonetheless, there are many subtleties when it comes to cluster the co-citation matrix. Since similarities can also be interpreted as a distance measure between documents, a wide variety of classic cluster methods can be applied. Nonetheless, the strongest restriction for clustering was demanding three documents as a minimum cluster size. Some relevant single documents observed could not be assigned to a corresponding cluster. This can be interpreted as a sign of emergent fields of research which could not yet establish their own research clusters.

- And how did the research field change over time?

To be frank, it was expected the field to have been widely disregarded within academic literature. On the one hand, Cleantech have attracted a lot of media attention. On the other hand, the abstraction level in describing certain technologies and their economic impact is rather low and could have discouraged researchers from exploring the field. The cocitation approach allowed me not only to gather relevant literature but also to classify emerging fields of research. By doing so, eleven clusters dealing with clean technologies were identified. Environmental demands for pure air and lower pollution led to certificate solutions which obviously inspired many researchers (Cluster 2). Air pollution issues seem to give noticeable impetus for transforming our economies very early. Two clusters, Clusters 1 and 6 describe the economic implications of this innovative sector on a very abstract level. Both clusters indicate the strong interest of economists to understand the interactions between economic growth and occurring environmental problems.

Not surprisingly, most clusters identified are related to renewable energies. Since 1998, many documents on renewable energies have been appearing regularly. The date corresponds to a governmental change to the SPD-Green-Party in 1998 which could have boosted the renewable energy sector. Interestingly, most researchers do not focus on a concrete technology when writing about renewable energies. Instead most concentrate on the rather abstract product "eco-electricity". Especially pricing and sales issues of eco-electricity in Germany seem to interest researchers.

During the first decade of the 2000s, most look at the renewable energy sector favorably, only a few researchers dare to view on the new sector more critically. However, newer research sees the Renewable Energy Act under critical light (see clusters 9 and 10).

Probably the most surprising result is the occurrence of the CCS cluster. It is the first cluster dealing with a concrete clean technology. This is astonishing since the technology is still in an experimental stage and has no practical applications yet. Probably the urgent need for solutions fighting climate change makes this technology more appealing than others.

In summary, one may conclude that high level research on certain technologies remains fragmented and scarce. Despite a strengthened public perception, I found only a few clusters dealing with sustainable technologies. The next chapter deals with sourcing risks of the industry, a topic which has not been addressed yet.

3. Sourcing risks and (financial) risk management options. The example of rare earth metals

"For the merchant, even honesty is a financial speculation."

—Charles Baudelaire

In recent years, the supply of rare earth metals has become a center of attention in many industries, especially in the automotive industry and in several Cleantech related applications. Many new technologies such as wind turbines, solar panels, batteries and highly efficient electric motors rely on a handful of Rare Earth Elements (REEs) which turn out to be critical input factors. Since rare earth minerals are produced almost exclusively in China, the sourcing of the required minerals has become increasingly difficult and expensive for Western-run firms. In general, China's domestic prices for rare earths differ significantly from export prices. It seems as if China systematically makes use of its nearly monopolistic position and market power in favor of its own industry location. When the Chinese government announced the tightening of export quotas, prices of many scarce minerals exploded in 2011. The prices for some minerals have multiplied as much as tenfold above 2010 price levels before dropping back down. According to a survey of different manufacturing industries conducted by PricewaterhouseCoopers (PwC), the sectors renewable energy (78%), automotive (64%) and energy & utilities (57%) have already experienced instability in supply and do not expect the situation to improve in the next five years (PricewaterhouseCoopers (PwC), 2011). For many strategic decision makers in firms, this leads to the question of what an appropriate strategy would be for reducing price and supply risks. The purpose of this chapter is to contribute to the discussion of mitigating price risks caused by strategic input factors. While this objective concerns the overall research question of how to counter supply and price risks most efficiently, this chapter particularly focuses on a more detailed question:

Can financial instruments, such as derivatives, help us to reduce price risks induced by holding positions of strategic rare earth metals?

At first sight, the question may appear unorthodox, but considering the history of derivative development, it becomes clear that derivatives are the first choice for mitigating price risks. Hedging uncertainties caused by natural conditions such as future harvests were among the first applications of derivatives. Early studies dealt with the development of general hedging theories in combination with futures and the theory of storage (Kaldor, 1939; Working, 1948, 1949; Johnson, 1960). During the oil price shock in 1973, Black and Scholes advanced their revolutionary work on option pricing, and researchers soon realized possible solutions for natural resources (Black and Scholes, 1973; Solow, 1974). Since then, derivatives have not only become one of the most powerful instruments in finance but also have helped to compensate for risky positions, especially in commodity markets (Brennan and Schwartz, 1985). However, with the work of Black and Scholes (1973), it has become increasingly relevant for researchers to adequately determine the stochastic price process of the underlying and the behavior of the convenience yield (Gibson and Schwartz, 1990). Observed price dynamics of commodities can often be characterized and modeled as mean reverting, heteroskedastic or discontinuous processes (Routledge et al., 2000; Merton, 1976; Askari and Krichene, 2008). In comparison with bond or equity markets, commodities show these effects due to seasonal and weather effects, observed inventories, changing demand, as well as storage and transportation costs (Brooks et al., 2011).

Coming back to the matter of strategic metal markets in general and rare earth metal markets in particular, it may be said that there are currently no cutting edge studies on the pricing dynamics or possible derivative solutions available. Due to its economic importance, experts in the field use aggregate supply and demand prospects in order to estimate price estimations. Despite their potentially high strategic value, those studies do not allow conclusions for asset price dynamics and hedging opportunities. A further goal of this chapter is to bridge the gap between the highly sophisticated theoretical considerations on commodity derivatives and the empirical demand for price dynamics research of certain strategic metal commodities. Therefore, Chapter 3 is structured as follows:

In Section 3.1, a general REE market overview is provided. After presenting the latest market developments and figures, I try to conclude some initial theoretical implications of the possible REE pricing behavior. Section 3.2 shows empirical time series and suggests possible stochastic processes. Based on the suggested processes, all parameters are estimated in order to model the time series. Based on the proposed processes, Section 3.4 derives optimal pricing and hedging strategies. As a framework, standard Arbitrage Pricing Theory (APT) is applied and known solutions for the proposed price processes will be considered. The last section, Section 3.5, discusses the results and provides alternative policy options for dealing with pricing risks.

3.1. Rare earth metal market description

This section outlines the market conditions for rare earth elements. It derives market characteristics which may helpful for us to understand the market conditions and to develop models which fit the delineated markets well. To begin with the analysis, I start with a look at historical market developments. The market evolution of the past is driven by technological needs and the development of new, rich deposits. The analysis shows traditional deposit areas which may be reactivated in the near future as well as market frictions which are mostly caused by a quasi-monopolistic market structure today.

From a geological perspective, the term "rare" tends to be misleading, as rare earth deposits occur frequently in the earth's crust. However, when REE deposits do not exceed adequate concentrations, they cannot be recovered economically. Since the 1950s, REE had usually been mined from several monazite deposits around the world. From 1965 onwards, REEs have mainly come from the rich Mountain Pass deposits in California, USA (Long et al., 2010). Accordingly, the time frame from 1965 to 1991 is also sometimes referred to as the Mountain Pass era (Tse, 2011). Most production facilities were located in the US around that time, and the Mountain Pass mine could strongly benefit from its strategic position. Around 1991, China began to exploit its own highly concentrated deposits and thereby undercut the prices of the American competitors. While Chinese production figures of REE increased more than four times from 1991 to 2007, prices fell to record lows by 2007 (Papp et al., 2008). The displacement competition has led to a Chinese dominance in the rare earth oxides market. By 2011, well over 90% of the world REE production is located in China; China is said to be in a quasi-monopolisitic situation today (Tse, 2011). Table 3.1 shows worldwide production figures and reserves for REEs in 2011. It can easily be seen that China does not only dominate the production but also possesses the richest ore reserves worldwide. The Chinese reserves are mostly concentrated in the Bayan Obo area, inner Mongolia.

In recent years, China has begun to use its market power by establishing trade barriers for rare earth metals. The restriction measures mainly include domestic production quotas, export quotas, and export duties. These are official measures to conserve domestic resources and protect the environment. The critics of the barriers believe, however, that China misuses its market power and forces foreign producers to relocate sensible parts of their supply chain to China. The market restrictions have led to WTO complaints and trade disputes. Japan, the EU, and the US filed disputes over rare earth metals in 2012 (WTO, 2012).

According to the current Chinese rare earth development plan, there is a production quota that limits production to 130–140 thousand metric tons of Rare Earth Oxide (REO) annually from 2009 to 2015. The export is limited to about 35 thousand metric tons (REO). Furthermore, only 20 domestic producers are allowed to export rare earth oxides. In addition to the mentioned export quotas, export duties were established in 2007. Export duty rates rose from 10% in 2007 to levels between 15 and 25% in 2011, depending on the metal (Tse, 2011). The increased prices also encouraged the implementation of other rare earth mine projects around the world for greater exploration efforts. Nonetheless, those projects need time and the market remains tense. Elements which are essential for production and prone to Chinese trade barriers are critically observed.

While the supply side was drastically tightened by the world's most important rare earth metal producer, China, the demand side has needed to react. In the short run, the Chinese measures have had a tremendous effect on the rare earth market and led to exploding prices. In the medium run, however, the market seems to have calmed down (Smith, 2012). The reasons for the current price situation are complex and open to speculation: it is likely that the most important driver for market dynamics is the Chinese economy. It seems that the Chinese economy is not fully independent of the financial crisis in 2008 and it has therefore created its own real estate bubble, that is, the economy has begun to grow at a slower rate. Other reasons may be found in producers abroad which have begun substituting scarce input factors and technologies. Some sources claim that the Chinese export barriers are difficult to enforce and thus carry little conviction. However, China increasingly punishes illegal mining activities and seems to gain back control over the market (Tse, 2011; Smith, 2012).

Before coming to an adequate description of rare earth price dynamics, I want to consider the market conditions from a more economical and theoretical point of view. The market can be characterized as a commodity market of scarce materials with a quasi-monopolistic market structure. Furthermore, the market seems to be underdeveloped in terms of free trade and market liquidity. With justification, it may be asked whether one can establish OTC contracts or standardized derivative products for markets with presumably strong frictions, such as forwards and options. To answer this complex question briefly, I want to give three counterarguments which suggest that the development of REE derivatives could make sense. The first comes directly from the derivative theory which shows, that under certain market conditions every derivative can be synthesized by a risk-free bond and the underlying stock when short selling is admitted. Then, all I would have to do is to replicate an adequate portfolio which promises the same pay-off as a comparable derivative instrument. The second argument comes from comparable markets with supply restrictions. Long term supply contracts are regularly used to reduce price risks. But eventually the question arises what a fair price of such a contract would be. The abovementioned arbitrage-free price can give a first clue here. The third argument is a prominent example in commodity pricing: crude oil and the

Organization of the Petroleum Exporting Countries (OPEC) cartel. Despite a fairly distorted spot market with fixed contingents, I observe a huge and quite well functioning derivative market. To conclude, it is difficult to determine the functioning of a derivative market by the microeconomic market structure only. In literature, I can mainly identify two lines of argumentation which may justify derivative markets that do not function. The first is the theory of storage and the second is the observation of incomplete markets. The theory of storage suggests that REE markets may face times of "backwardation", that is, future delivery prices decline with time-to-delivery. It might be more important to physically possess rare earth metals instead of holding abstract "paper" compensations. The benefit expressed as a rate is frequently referred to as "convenience yield" (Kaldor, 1939). Markets in backwardation tend to signal shortages in the market and more importantly indicate that physical delivery could be difficult (Fama and French, 1987, 1988). However, shortages can be observed in several commodity markets and may not be grave enough to justify dysfunctional derivative markets. An even stronger argument is the level of market incompleteness. Perfect, arbitrage-free replication strategies as proposed by Black and Scholes (1973) for options and Black (1976) for futures are regularly impossible in frictioned markets with institutional rigidities (Bertsimas et al., 1997). Despite pricing and hedging in incomplete markets being much more challenging, under certain conditions hedging strategies could still make sense. However, the key to possible pricing and hedging strategies is the knowledge of the underlying stochastic price processes. The next section uses REE spot price time series and proposes empirically founded stochastic processes. The analysis is followed by an estimation of process parameters and the deviation of apt valuation formulas.

Table 3.1.: Worldwide mine production and reserves in metric tonnes (REO), 2011 (USGS, 2012).

Country/Region	Mine Production	Reserves
US	-	13.000.000
Australia	-	16.000.000
Brazil	550	48.000
China	130.000	55.000.000
India	3.000	3.100.000
Malysia	30	30.000
Commonwealth	NA	19.000.000
Other	NA	22.000.000

3.2. Valuation principles for rare earth commodity derivatives

After describing the conditions in the REE market, the REE spot price time series should be analyzed in greater detail. In the first subsection, I describe the time series and the proposed stochastic processes. After this, empirically founded process models are proposed and the respective parameters are estimated. The second subsection, derives equivalent processes in the world of a risk-neutral investor, and the foundation for pricing and hedging strategies with derivatives is laid.

3.2.1. Available data

Until today, the rare earth metal markets could not yet establish a formal market with a standardized commodity exchange. Thus, rare earth metal prices cannot simply be obtained by standard financial data providers, but by special trade and data agencies which are primarily located in China or have strong business connections with Chinese counterparts. The data provider Bloomberg, for instance, mainly obtains its rare earth metal input data from Steelhome, China. I had the opportunity to compare several time series covered by Bloomberg with those provided by Metal-Pages[8] and found out that Metal-Pages price data usually reacts earlier when new information enters the market. Therefore, rare earth metal prices as covered by Metal-Pages will be used for further analysis. The prices are usually calculated on th e basis of extensive discussions with producers, consumers and traders, and do also reflect actual trades.

Rare earth metal prices depend strongly on the intended destination of use (inside or outside of China), the quality and purity of the metals and oxides offered and the International Commercial Terms (Incoterms). For western firms, Free On Board (FOB) China prices in US Dollar are the most relevant entity. The FOB China data set consists of 27 rare earth metal price time-series. The majority of these series ranges from 01/02/2002 to 11/29/2012 which roughly covers more than ten years. Some time-series have not been published before 2007 and two further time series have been added in late 2012. An overview of all data sets can be found in the Appendix A. In Figure 3.1, the typical price development of Lanthanum metal with a minimum purity of 99% is presented.

Without proof, it can be easily seen in Appendix A that all observed rare earth metals and oxides are highly correlated with each other. This is not a surprising observation, since most rare earth metals are co-products extracted from the same

8 Metal-Pages is a special portal and data provider in the field of metals and especially rare earth metals. Available: www.metal-pages.com.

Figure 3.1.: Path of Lanthanum metal FOB China min 99% a) spot prices and b) corresponding log-returns from January 2002 till November 2012.

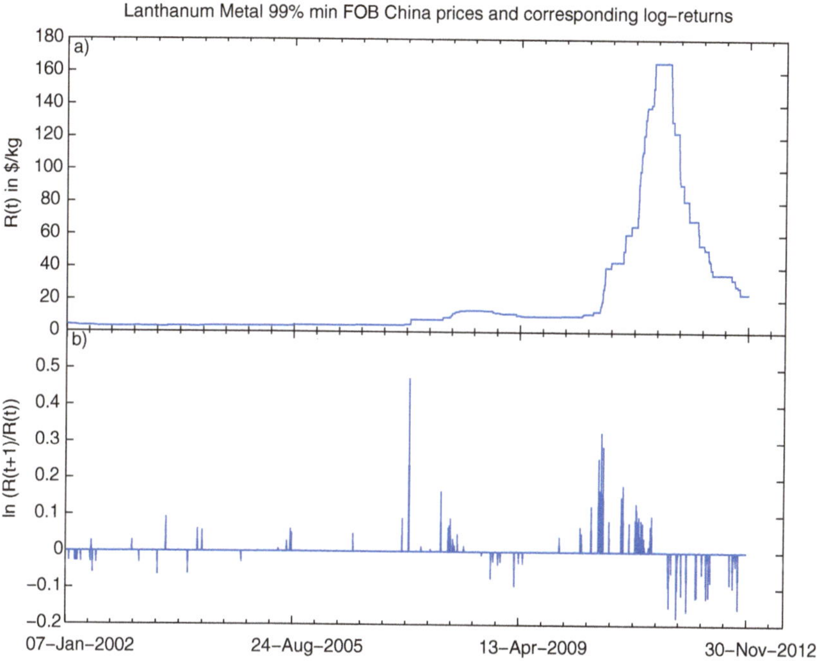

ores while the demand side remains constantly at high levels. All time-series show a relatively calm era between 2002 and 2009 with relative low volatility. By the turn of the year 2009/2010 prices for every REO began to soar until end of July 2011 and partially reached more than the tenfold level compared to 2009 prices. During that time, not only the returns went up but the trade intensity began to rise as well. Since then, the market obviously began to calm down again. The observed downward trend is of comparable size as the previous upward trend which actually counteracts Chinese price stabilization efforts. The data set also reveals some interesting characteristics which will be needed for the determination of apt stochastic process models in the following section. From Figure 3.1, it can be observed that there are many days with no price movement and sudden discontinuous price shifts when new information enters the market. It is interesting to note that there is obviously no diffusion or continuous drift in the price dynamics, there are no signs of Brownian motion. The next section assumes the here mentioned data set as a possible stock underlying and proposes corresponding models for the stochastic price processes.

3.3. Stochastic price process models

In this section, two stochastic process models for the pricing of rare earth derivatives are presented.

Model 1

For the first model, I simply assume a compound Poisson process which can deal with occasional discontinuous breaks of random height. For this very basic approach the Poisson process and the compound Poisson process id defined first.

The Poisson process (sometimes also referred to as the counting process) counts π occurring events (jumps) during a certain time interval h. A counting process $X(t)$ is a homogenous Poisson process $\pi_{t,\lambda}$ with intensity function λ and time t, if the following conditions are met (Merton, 1976):

1. At time 0 there is no count, hence $\pi(0) = 0$
2. The number of events in two disjoint time intervals are independent, that is the process increments are independent.
3. The probability of one event during an interval of length h is $P((X(t+h) - X(t) = 1) = \lambda h + O(h)$
4. The probability of more than one event during an interval of length h is $P((X + h) - X(t) \geq 2) = O(h)$ where $O(h)$ is the asymptotic order symbol and the limit $O(h) = \lim_{h \to 0} \frac{O(h)}{h} = 0$.

Based on this definition, the compound Poisson process is defined as

$$\pi_t^Y = \sum_{n=1}^{\pi_t} Y_n \tag{3.1}$$

Model 2

Figure 3.1 also reveals some sudden shifts in the pricing behavior of the delineated time series, namely at the turn of the year 2009/2010 and after 08/01/2011. On first sight it cannot be determined whether the shifts were caused by changed jump sizes Y or changed intensity rates λ. The observed states between the shifts appear to be persistent for a while. In addition to equation 3.1, I therefore assume that the jump sizes and intensities λ may vary stochastically with time and remain at the same state for a while. A straightforward assumption to model the stochastic parameter change would be the application of regime switching processes as proposed by Hamilton (1989) and applied by several authors for financial time series in the field of option pricing (Bollen, 1998; Duan et al., 2002).

For the second model, I therefore write an equation similar to equation 3.1 for each regime state s_t.[9]

$$dR = \mu_{st} R dt + Rd\left(\pi^{Y}_{st,t}\right), \text{ with states } st = 1,\ldots,k \tag{3.2}$$

Furthermore, I assert that the transition between regimes can be described by a first-order Markov chain. That is, the transition probability between state i and state j can be written as transition matrix P_{ij}

$$Prob[s_t = j | s_{t-1} = i] = P_{ij} \tag{3.3}$$

3.3.1. Parameter estimation

In this subsection the parameter estimation for the above proposed price process models is discussed. I explain the estimation procedu re on the basis of the example of Lanthanum Metal time series. The procedure was performed for all given rare earth time series. For the results, I refer to Appendix B.

The models proposed above mainly rely on two description parameters, the jump intensities λ and the jump sizes of the log-normal returns Y. The first model proposes a homogeneous compound Poisson process.[10] Since the intensity rate λ is homogeneous, I make use of the maximum likelihood estimator which is simply the average of all counted jumps within a given time period.

$$\hat{\lambda}_{ML} = \frac{1}{n} \sum_{i=1}^{n} (Y_i) \tag{3.4}$$

The jump sizes Y can estimated parametrically or with non-parametric kernel density distributions. In the case of Lanthanum metal, I measure an overall intensity rate $\hat{\lambda} = 0.032$ for the time period between January 7th 2002 and November 30th 2012. Furthermore, I fit a distribution to the log-normal jump-sizes Y. A Q-Q-plot indicates that a t-location-scaled distribution with three parameters, $\mu = 0.012$, $\sigma = 0.080$ and $v = 4.578$, fits well to the jump sizes and reproduces the heavy tail characteristic of the returns. For more accuracy, a non-parametric kernel distribution can be fitted. In Figure 3.2, I show the Q-Q-plots of the Lanthanum Metal log-returns for the best parametric fit and the non-parametric

---

9 Note that a change in intensities $d\pi(\lambda_k)$ effect the average drift μk accordingly.

10 In the Appendix, I actually show that the intensity rate is inhomogeneous. However, homogeneous intensity rates can be assumed for longer periods of time. Furthermore, I show that the intensity rates are not the prevalent for the valuation in a risk-free world.

kernel distribution. Summarizing the estimation efforts for model 1, I conclude that the model can be fitted well to the empiric data. The model obviously remains valid for longer periods of time but eventually I observe sudden shifts, both in intensity rates and jump sizes.

Therefore, the second model proposes that the price process is governed by a regime switching compound Poisson process. In this case, I am not only interested in parameters for the intensity rate λ and jump sizes θ but also for the transition probabilities P_{ij} between several states. In order to calculate these, literature

Figure 3.2.: Quantile plot of Lanthanum metal log-returns. The left hand illustration shows the fitted t-location-scaled distribution with parameters μ, σ and ν. This can be compared to the more accurate non-parametric kernel-distribution on the right.

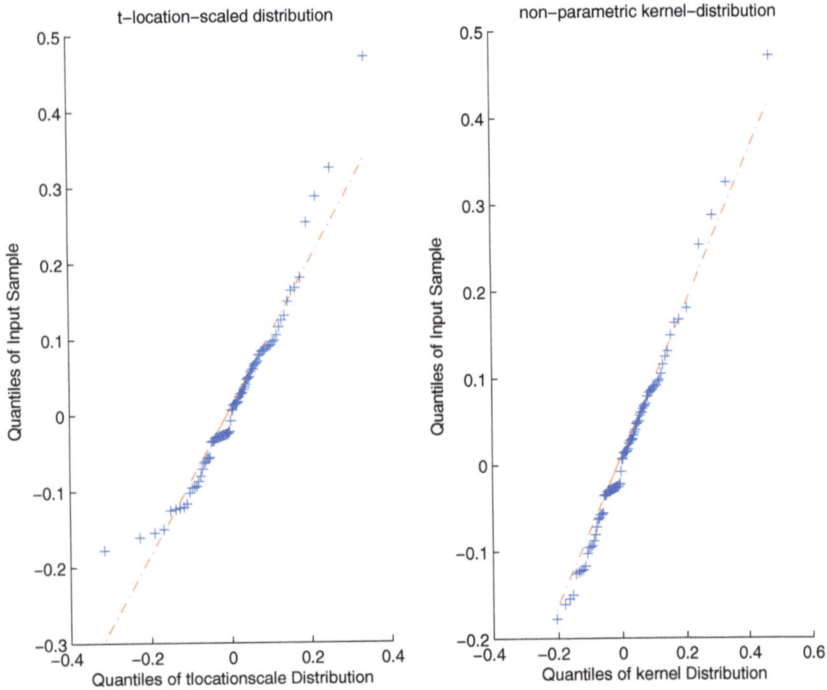

proposes at least two standard approaches (Hamilton, 1989; Ryd'en, 2008). One is the more powerful[11] but also more complicated estimation via Markov Chain Monte Carlo (MCMC) methods (Green, 1995), the other approach is estimation

11 MCMC also allows to calculate estimation errors.

via the Expectation-Maximization-Algorithm (EM-Algorithm) and instances derived from it (Dempster et al., 1977; Hamilton, 1989). For reasons of simplicity, the EM-Algorithm for the estimation of the transition matrices is applied to my time series.

From Figure 3.1 and Appendix B.1, I deduce for all time series that the spot prices behave relatively stable until 12-31-2009 and seem to switch to another state afterwards. Thus, I estimate the intensities $\lambda_{1,2}$ and the jump kernels (alternatively parameters) $\theta_{1,2}$ before and after the date marking 12-31-2009. The second observed regime switch occurs around 07-31-2011 and it can be handled the same way.

The estimation results are shown in Figure 3.3. The probabilities for being in state $s_t = 1, 2, 3$ in dependence of the corresponding parameters $\lambda_1, \theta_1; \lambda_2, \theta_2; \lambda_3, \theta_3$ are calculated and it can be observed that each state was very stable for a fairly long time period. For the above example, Lanthanum metal, the following values for the transition matrix $\widehat{P}_{ij}$ were calculated by the EM-Algorithm

$$\widehat{P}_{ij} = \begin{bmatrix} 0.9984 & 4.98 * 10^{-4} & 0.0011 \\ 8.35 * 10^{-4} & 0.9992 & 9.83 * 10^{-25} \\ 0.003 & 2.35 * 10^{-37} & 0.997 \end{bmatrix} \tag{3.5}$$

High probabilities near 1 on the diagonal of the transition matrix P_{ij} indicate that each state can be characterized as persistent. The results suggest that the market can retrospectively characterized by three phases: a “normal” phase 1, a “bullish” phase 2 and a “bearish” phase 3. The first phase until 2010, can be described as calm in terms of trade intensity and almost “normal” distributed log-returns during that time can be observed. The second phase, between 01-01-2010 and 07-31-2011, can be described as a bullish market with much higher intensity rates and higher returns. Then eventually the third phase can be described as a period with still high intensity rates but much more negative returns. The estimation results also suggest that Model 1 is still accurate and remains valid within each state but cannot reproduce the dynamic change of the intensity and return parameters. Model 2, however, takes the observed dynamics into account and can theoretically be adjusted to all of the 27 observed time series.

Based on the results obtained so far, the basic pricing principles for the suggested price dynamics should be derived.

Figure 3.3.: Regime probabilities as a function of time depending from the jump distribution parameters θ and intensity parameters λ. Lanthanum Metal from Jan 2002 till November 2012 is considered.

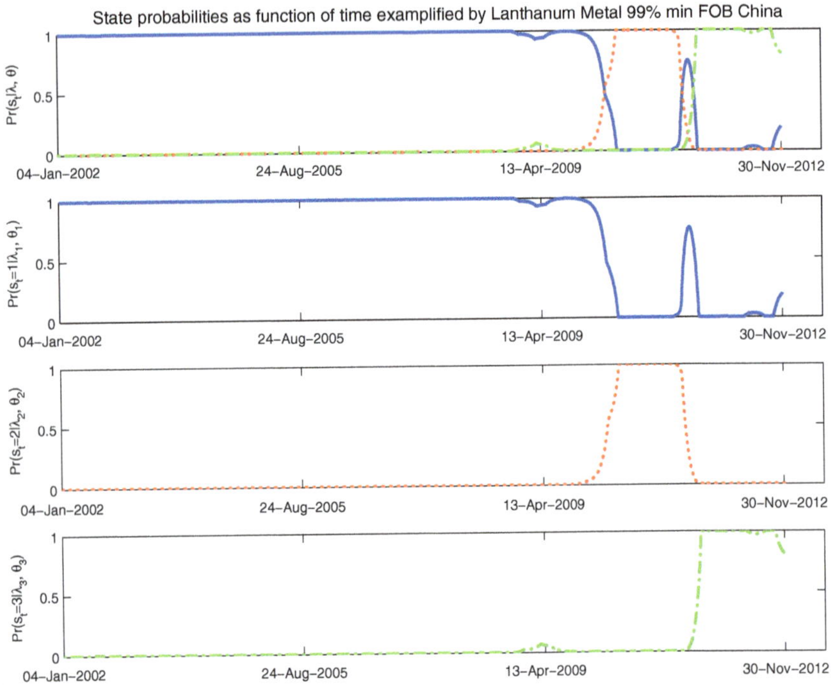

3.3.2. Valuation via risk-neutral measures

Up to this point, stochastic price processes with a well fit have been determined for all time series (equations 3.1, 3.2 and 3.3). In this subsection, I would like to answer the question of whether these stochastic price processes can be used for derivative valuation purposes. To do this, the well-known standard Arbitrage Pricing Theory (APT) framework is applied. The basic idea in APT is to build a self-financing portfolio of a risk-free bond and the underlying stock and try to replicate the returns of the derivative. The no-arbitrage condition in this framework leads to the concept of the risk-neutral investor, which means that derivatives should be priced as if they were in an imaginary risk-neutral world and therefore must offer the same return as the risk-free bond in order to avoid arbitrage strategies.

For valuation purposes, it makes sense to analyze the behavior of the proposed compound Poisson processes in a risk-free world. As a result, the most important scientific valuation principles are briefly reiterated. A more detailed look at geometric Poisson motion can be found in Aase (1988). For the case discussed here, I refer to Shreve (2004); Elliott and Kopp (1990) and Cheang and Chiarella (2008) for a portfolio of multiple discontinuous stocks.

To begin, it is convenient to decompose the compound Poisson process into M counting processes with equal jump sizes y_m (Shreve, 2004). I may write

$$\pi_t^Y = \sum_{m=1}^{M} y_m \pi_m(t) \tag{3.6}$$

and

$$\pi_t = \sum_{n=1}^{M} \pi_m(t) \tag{3.7}$$

It is important to note, that π_t^Y as a risk factor may have drift if $\mathbb{E}(Y_i) \neq 0$. Hence, the expected constant drift term needs to be subtracted to derive a martingale property. The expected drift can be denoted as

$$\beta = \mathbb{E}(Y_i) = \sum_{m=1}^{M} y_m p(y_m) = \frac{1}{\lambda} \sum_{m=1}^{M} y_m \lambda_m \tag{3.8}$$

and the process

$$\pi_t^Y - \mathbb{E}\left(\pi_t^Y\right) = \pi_t^Y - \mathbb{E}(Y_i)\lambda t = \pi_t^Y - t\sum_{m=1}^{M} y_m \lambda_m \tag{3.9}$$

is a martingale. Based on these definitions, the two suggested price process models are analyzed.

For the first model, I assumed that all considered rare earth metals follow a Compound Poisson Process with constant drift μ, an underlying process which was first proposed by Cox and Ross (1976). With $\hat{\mu} = \mu + \beta\lambda t$, equation 3.1 can be written equivalently in a differential form as[12]

$$dR = \hat{\mu} R dt + R d\left(\pi_t^Y - \beta\lambda t\right) \tag{3.10}$$

Here, the process is defined with a constant drift μ and martingale Poisson factor. The solution to equation 3.10 is

$$R(t) = R_0 exp\{(\mu - \beta\,\lambda)t\} \prod_{1}^{\pi_t^Y} (Y_i + 1) \tag{3.11}$$

---

12 Actually, no constant drift term is observed. Hence, $\hat{\mu} = \beta\lambda t$.

In the next step, I would like to consider equation 3.10 in a risk-neutral world. That is the change of the real world probability space $(\Omega, \mathbb{F}, \mathbb{P})$ to the probability space of the risk-neutral world $(\Omega, \mathbb{F}, \mathbb{Q})$, where $\mathbb{Q}$ is the probability assignment of the risk neutral world.

Let $m_\pi(z)$ be the moment generating function for the compound Poisson process π and $m_Y(z)$ the moment generating function for the random jump sizes Y_i, then

$$mY(z) = \mathbb{E}(e^{zY_i}) \tag{3.12}$$

and

$$m_{\pi_t^Y}(z) = \mathbb{E}(e^{z\pi_t^Y}) = exp\left\{\lambda t\left(\sum_{m=1}^{M} p(y_m)e^{zym} - 1\right)\right\} \tag{3.13}$$

Furthermore, let $Z(t)$ be the Radon-Nikodym derivative of the compound Poisson process, that is under the risk-neutral probability measure $\mathbb{Q}$, the intensity rate changes from λ to $\lambda\mathbb{Q}$ when exactly $\pi(t)$ jumps have occurred by t. For all m decomposed Poisson processes, the Radon-Nikodym derivative is defined as

$$\frac{d\mathbb{Q}}{d\mathbb{P}} = Z(t) = \prod_{1}^{M} e^{(\lambda_m - \lambda_{m\mathbb{Q}})t}\left(\frac{\lambda_{m\mathbb{Q}}}{\lambda_m}\right)^{\pi(t)} \tag{3.14}$$

The goal is to obtain the expected value of the compound Poisson process under $\mathbb{Q}$ which is

$$\mathbb{E}_{\mathbb{Q}}(e^{z\pi_t^Y}) = \mathbb{E}(e^{z\pi_t} Z(t)) = exp\left\{\lambda_{\mathbb{Q}} t\left(\sum_{m=1}^{M} p_{\mathbb{Q}}(y_m)e^{zy_m} - 1\right)\right\} \tag{3.15}$$

When comparing equations 3.13 and 3.15, Shreve (2004) shows that compound Poisson processes do change the intensity from λ_m to $\lambda_{m\mathbb{Q}}$ and the probability jump size distribution from $p(y_m)$ to $p_{\mathbb{Q}}(y_m)$ under $\mathbb{Q}$.

In summary, it can be statet th at under Q, the compound Poisson process π_t^Y has independent jumps with sizes Yi and distribution $p_{\mathbb{Q}ym} = \lambda_{\mathbb{Q}} / \lambda$. The intensity rate is equal to the sum of all m intensity rates $\lambda_{\mathbb{Q}} = \sum_1^M \lambda_{m\mathbb{Q}}$.

Till this point, a rule has been derived which allows to change probability measures for compound Poisson processes. In order to avoid arbitrage, it must be required that the return of the compound Poisson process must be equal to the return of a risk-free bond r. That is

$$dR = rR(t)dt + R(t^-)d(\pi_t^Y(t) - \beta_{\mathbb{Q}}\lambda_{\mathbb{Q}}t) \tag{3.16}$$

or equivalently

$$\mu - \beta\lambda = r - \beta_Q \lambda_Q \tag{3.17}$$

Recalling equation 3.8, equation 3.17 can be rewritten and the market price of risk equation for the problem can be derived as

$$\mu - r = \beta\lambda - \beta_Q \lambda_Q = \sum_{1}^{m} (\lambda_m - \lambda_{mQ}) y_m \tag{3.18}$$

According to Shreve (2004) the solution to equation 3.16 is

$$R(t) = R_0 exp\{(r - \beta_Q \lambda_Q)t \prod_{1}^{\pi_t^Y} (Y_i + 1)\} \tag{3.19}$$

With the equations derived so far I have all the tools at hand to price derivatives determined by compound Poisson processes with arbitrary jump sizes. With Monte-Carlo simulation of equation 3.19 it is directly possible to calculate the contingent claim paths. However, equation 3.18 makes absolutely clear, that there is a wide set of solutions available to fulfill the equation – the market is incomplete. For concrete pricing or hedging, reasonable further restrictions are needed. Section 3.5 discusses known approaches for further restrictions in incomplete markets.

3.4. Excursus: Known closed-form solution for constant jump sizes

For stochastic differential equation (3.1), there exists an analytic solution if the jump sizes are known and constant (Cox and Ross, 1976; Cox et al.,1979). I think the formula is very useful for the calibration of numerical solutions, such as Monte-Carlo methods. Figure 3.4 summarizes the binomial situation for the underlying. In a binomial setting, the change of the rare earth metal price dR can be modeled as a composition of a constant drift term μ in infinitesimal time dt and the Poisson jump factor. In time span dt, the Poisson jump factor jumps upwards u with probability λdt and remains down d with probability $(1 - \lambda)dt$.

Figure 3.4.: Binomial tree for jump process equation (3.1) as proposed by Cox and Ross (1976).

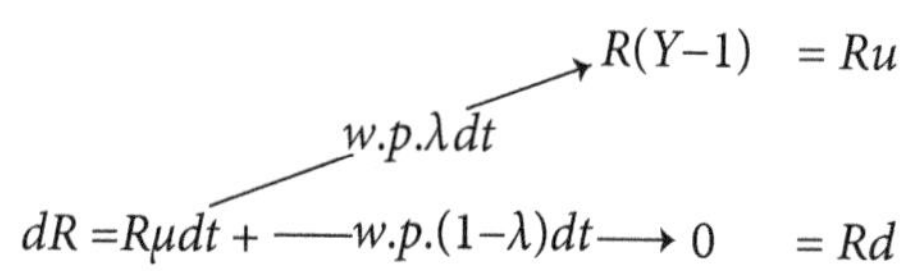

In this setting, the binomial tree can be solved for n discrete time spans and the general solution for an European type call option is given by

$$C = \lim_{n\to\infty} R(0)\sum_{i=b}^{n} b(i \mid p_n') - Kr^{-n}\sum_{i=b}^{n} b(i \mid p_n) \tag{3.20}$$

where $b = ln(K/Sd^n)/ln(u/d)$, $p = (r - d)/(u - d)$ and $p' = (u/r)p$. To adopt this general binomial solution to the solution of a Poisson jump process, I substitute $u = Y$, the observed jumps, $d = e^{\mu(t/n)}$, a constant drift, $q = \lambda(t/n)$, the jump probability and let $n \to \infty$ (Cox et al., 1979). In combination with Poisson limit theorem (Page and Sanders, 1986) I derive

$$np_n = \frac{n(e^{\frac{rT}{n}} - e^{\frac{\mu T}{n}})}{Y - e^{\frac{\mu T}{n}}} \to \lim_{n\to\infty} \frac{n(e^{\frac{rT}{n}} - e^{\frac{\mu T}{n}})}{Y - e^{\frac{\mu T}{n}}} = \frac{(r-\mu)T}{Y-1} \tag{3.21}$$

And for the call option:

$$C = R(0)\sum_{i=b}^{\infty} \frac{\left[\frac{Y(r-\mu)T}{Y-1}\right]^i e^{(-\frac{\mu(r-\mu)T}{Y-1})}}{i!} - Ke^{-rT}\sum_{i=b}^{\infty} \frac{\left[\frac{(r-\mu)T}{Y-1}\right]^i e^{(-\frac{(r-\mu)T}{Y-1})}}{i!} \tag{3.22}$$

where $b = min\left(\frac{ln\left(K/R(0)\right) - \mu t}{ln\, u}\right) \Lambda\, b \in \mathbb{N}$.

The solution (3.22) can then be expressed as (Cox et al., 1979)

$$C = R(0)\psi(a,b) - Ke^{-rT}\psi(a,\frac{b}{Y}) \tag{3.23}$$

where $a = \frac{(r-\mu)YT}{Y-1}$, $b = min(\frac{ln(K/R(0)) - \mu t}{lnY}) \wedge b \in \mathbb{N}$ and $\psi = \sum_{i=b}^{\infty} \frac{e^{-a}a^i}{i!}$.[13]

This is a Black-Scholes type solution which calculates Poisson factors ψ instead of the well-known normal distributed factors and considers jump sizes Y instead of volatility σ.[14] Note that in this case the market can be characterized as complete.

---

13 b ensures that only paths with positive payoffs (in the money) are counted.

14 Note that ψ is nothing but the Poisson Cumulative Distribution Function (CDF) from b to infinity and can easily be calculated $F(\infty) - F(b)$.

3.5. Pricing and hedging rare earth metals: An incomplete market

It is a very well-known fact that discontinuous stochastic processes may not necessarily though likely lead to incomplete markets. In this section, I would like to describe more precisely what economic theory defines as an incomplete market and discuss implications for the rare earth metal markets.

According to Harrison and Pliska (1981) it is not the discontinuities of stock prices that cause incompleteness in a market but the impossibility for investors to attain all contingent claims at a certain point of time. To be more precise, one can define that a contingent claim is a measurable random variable H_t. A contingent claim can be reproduced if there is a self-financing trade strategy ϕ which allows the value of the portfolio to be equal to the contingent claim $V_T(\phi)$. A market is complete if there is a replicating portfolio for every contingent claim H_t, otherwise the market is defined as incomplete. From equation 3.18 it can be easily seen that a compound Poisson process with random, identically distributed jump sizes leads to incomplete markets (Shreve, 2004).

Incomplete markets may be caused by illiquidities, market closures, monopolies and other macro market distortions which in turn lead to the observed price discontinuities (Bertsimas et al., 1997; Merton, 1998). In this case, APT pricing approaches deliver a price range for the derivative instead of a single non-arbitrage price. As the main consequence, investors cannot completely mitigate price risks with derivatives. Incomplete markets are therefore also referred to as markets with weak hedging opportunities. Nonetheless, I would like to shortly mention some popular approaches which may help to restrict the set of solutions to equation 3.18 and may be used for pricing and hedging opportunities.

The first approach is the assumption that an "average" log-return jump size can be estimated from the data. Then, assuming that $P(Y = \mathbb{E}(\theta)) = 1$, equation 3.18 reveals only one solution and equation 3.19 can be reduced to the equations analyzed by Cox and Ross (1976) (see excursus in Section 3.4). Despite Model 2 introduces a regime-switching process, it seems to be beneficial in comparison to Model 1, because it reduces the codomain of possible jump sizes for each state. However, there is no analytic solution for the regime-switching process and solutions must be calculated via Monte Carlo methods. To do this, I must assume that the transitions probabilities P_{ij} and associated persistence times for each state are the same in probability spaces $\mathbb{P}$ and $\mathbb{Q}$.

Besides the "average jump size hedge" methods, I would also like to shortly present some alternative approaches which help to reduce the solution space in equation 3.18. An approach with strong theoretical appeal is the

Expected Utility Maximization (EUM) according to Davis (1998). The idea is to make use of a concave utility function U in order to define the risk which is preferred by market participants. The goal of each participant is to maximize the utility function for a given trade strategy under the real probability measure $\mathbb{P}$. Hence, every individual maximizes their utility function $sup\mathbb{E}(U\,(V_t(\phi)))$ (Davis, 1998). However, the method demands the assessment of personal utility functions for each market participant which makes the procedure rather infeasible in practice. Solutions can be found with dynamic programming or duality approaches.

A third class of approaches can be summarized as quadratic hedging strategies (Duffie and Richardson, 1991; Schweizer, 1992, 1995). These strategies pursue the idea of finding hedges ϕ which at least allow to minimize the risk of holding unattainable contingent claims H. The hedging strategy ϕ earns amounts ϕ_t at price R_t. The cumulated gain of such a strategy until time T is (Schweizer, 1992):

$$G(\phi)T = \int_0^T \phi_t dR_t \tag{3.24}$$

Furthermore, I have contingent claims H, as defined above. The hedger faces constant initial cost L. In this case, the goal in variance-optimal hedging is to minimize the hedging error, which is the quadratic term (Schweizer, 1992):

$$min_\phi \mathbb{E}\left[(H + L - G(\phi)_T{}^2\right] \tag{3.25}$$

The approach globally optimizes a hedging portfolio in terms of a minimum variance martingale measure. On first sight it seems to be disturbing that gains G are punished but for simplicity reasons and the given insight into the problem, the approach is certainly appealing (Hubalek et al., 2006). I conclude that despite pricing and hedging in incomplete markets having become topic of debate for most financial mathematicians, the solutions for pure jump processes with random jump distribution remain unsatisfying.

3.6. Discussion of Chapter 3

In this last section of Chapter 3, I would like to discuss the results and point out alternative policy options for rare earth metal price risks. Therefore, I briefly recapitulate the work done so far. First, the stochastic price processes, which can be observed in rare earth FOB China spot markets, have been analyzed. While the case of jumps and corresponding Poisson processes is widely discussed in the theory, I have found no empirically discussed cases with pure jump characteristics in commodity markets. In most cases, researchers focus on processes which may be a mixture of diffusion and jump risk factors. The reason may be twofold: to my mind, pure jump processes may be a strong indication of a malfunctioning market per se: low liquidity, state interventions and captive markets may easily lead to discontinuous trading conditions. The second reason may be a more pragmatic approach: diffusion risks may be priced easily while jump risks regularly cannot.

Nonetheless, the corresponding stochastic price processes have been estimated and it is shown that the processes can be approximated with compound Poisson processes with homogeneous jump intensities and jump sizes which can be derived from empirically determined distributions. Sudden shifts in intensity rates indicate possibly Chinese market interventions. Unfortunately, it can be easily seen that the equation for the market price of risk has several solutions which correspond to unknown jump sizes and there is only one security available to attend possible contingent claims. In short, I diagnose an incomplete market. This has led to the discussion of some well-known hedging strategies, which may help to attenuate possible price risks. Nonetheless, the promising idea of hedging rare earth price risks remains strongly limited by incomplete market conditions. I therefore conclude that there is no satisfying derivatives solution which could promise a substantial reduction of price risks.

However, this assessment could change: old mines in Mountain Pass, California and new deposits in Australia, Greenland and the Middle East are successively going to be utilized. With new market participants entering the market, the observed stochastic processes will likely change toward a higher "degree" of market completeness. It may be possible that the market develops price processes with combined diffusion and jump proportions. While the diffusion factor can be hedged, the jump factors will remain. This work may enable a better understanding of the empirical jump characteristics. I see that jumps are mainly driven by changes in intensity and less by a change of jump sizes over time. This observation eases pricing and hedging in jump diffusion environments, since intensities do not influence the risk-neutral martingale measure under Q, while changes in jump size do.

Against the backdrop of weak hedging opportunities, the question of whether economists may offer alternative options in order to pose possible price risks arises. One simple and natural way would be the participation in new mining activities which could directly benefit from REE price gains. However, all equity investments pose the risk of being rather engaged in new business risks than in hedging opportunities for price risks. Thus, I see a substantial sourcing risk for many industries and especially for the young E-mobility industry. For firms involved, I identify mainly three alternative strategies to address price risks:

1. Substitution and recycling
   The first of those options - and probably the most important one is the avoidance of rare earth metals in engineering designs. Various products can be designed differently to help avoid supply problems. While many REEs can be substituted, others cannot. Those could probably be recycled. Unfortunately, actual recycling technologies are still relatively undeveloped and recovery rates are low. This is especially the case if waste disposal is not carried out by the initial manufacturer. Knowledge of the exact composition of products is essential for efficient recycling processes.
2. Creating a stockpile
   Creating a stockpile is not only an option for state policies but also for privately run companies. Interestingly, companies tend to collaborate to increase market power when negotiating with Chinese suppliers. For example, German high-tech industries have formed a commodity alliance for joint purchasing initiatives and joint REE coverage.[15]
3. Relocation of production and outsourcing of value chains
   Permanent price risks may also lead to outsourcing decisions on China. Rare earth products made in China are exempt from export duties and enjoy price advantages. Despite a fairly competitive environment in China, foreign investors may fear the circulation of internal knowledge and possible lock-in effects following outsourcing decisions.

In addition, there are several state-driven policy options which could help to reduce sourcing risks. I therefore refer to Humphries (2012).

Before coming to the conclusion, some methodological shortcomings of Chapter 3 should be discussed briefly. First of all, the data available is rather weak and questionable. There are mainly three data providers which publish FOB China spot prices. Since there is no official stock exchange, the data observed comes directly from stock traders and is of questionable quality. In addition, I could not observe Over The Counter (OTC) forward prices which could help to estimate not only price risk factors but also stochastic convenience yields which proved to be useful in other frictioned commodity markets. In this chapter, two pricing models are proposed. All needed pricing formulas for the first model have been revived, a complete derivation for the second is missing. But the regime-switch model can cope with this problem: The equations remain valid but it must be assumed that the transition probabilities remain the same in different probability spaces. While this solution will work numerically, it does

15 See http://rohstoffallianz.com/ for further information.

not provide a closed form solution and it is therefore difficult to determine a complete set of valid solutions.

In summary, I conclude that derivatives are not expected to be an efficient solution to the apparent price risks in the REE market. I see the reason being in market incompleteness which can be associated with poor hedging opportunities by definition. The study raises questions regarding how to easily measure the "degree of market incompleteness" (Bertsimas et al., 1997), which is also the most important measure for the hedge ability. With new market participants entering the REE market, it would be extremely useful to know how spot and forward prices are effected and to what extent the hedge ability is increased. Currently, firms are well advised to mitigate all REE market dependencies.

4. E-mobility and capital market perception

> "For a successful technology, reality must take precedence over public relations, for Nature cannot be fooled."
>
> —Richard Feynman

From an automotive firm's perspective, the concept of E-mobility raises several strategic management decisions. The development of electric vehicles and the necessary integration of technologies is very costly and could quickly turn out to be a very risky venture for car makers. This is even truer if new series models cannot be offered at competitive prices or design and subsequently fail in the market. Otherwise, it could be as risky for car makers to ignore the trend and to be incrementally driven out of business. This ambiguity leads to the management question of how to gain competence in the field and access new markets without going to much into risk. That applies all the more for the emerging E-mobility industry, as many fields of technological knowledge originate from other, external industries, such as the chemical industry (Kasperk and Drauz, 2013; Proff and Proff, 2013).

In the automotive industry, partnership and cooperation is an usual, well-known and frequently applied strategy. Especially the idea of accessing new, broader markets but also the issue of developing (technological) competence has led to a huge number of cooperations in the industry (Wallentowitz et al., 2010). In the context of diminishing fossil fuels and climate change, the industry and its cooperative structures is facing major changes (Hanselka and Jöckel, 2010; Lienkamp, 2012). The observation of uncertainty in the market leads to two main questions which should be, if only partly, addressed in this chapter.

The first question I would like to address is whether capital markets positively evaluate partnerships and therefore reward all efforts to integrate technological knowledge gains in the area of E-mobility by cooperation. The question of capital market perception regularly arises for many industries as it expresses value generation of management decisions. However, as mentioned in the introduction, Proff and Proff (2013) conclude that no capital market oriented study with clear results has been published yet.

Other industries suggest that the simplest another way to win the race of technology leadership would probably be the direct acquisition of relevant companies with respective technological knowledge. Observing the E-mobility market from the start of this millennium, however, I do not observe much M&A activity from OEMs which might suggest that a race for being the first-mover is going on. The same applies for other typical capital market activity, such as IPOs or possible Seasoned Equity Offerings (SEOs) with a clear E-mobility focus. Instead, it is collaborative activities that have substantially grown since then and make first capital market related evaluations possible.

Proponents of positive collaborative wealth effects would refer to the idea of value creation effects and possible positive prospects for the commercialization a new technology. Former event studies which consider different industries suggest that alliances and joint ventures may lead to positive abnormal returns (McConnell and Nantell, 1985; Koh and Venkatraman, 1991; Chan et al., 1997). On the other hand, Proff and Proff (2013) also mention that many cooperations in the automotive industry often failed in the past which leads to literature of possible cooperation failures and rather suggest value destruction by cooperation. I therefore refer to Kale et al. (2000); Ariño et al. (2001); Danzon et al. (2005) for problems of moral hazard, knowledge leakage and potential competitive disadvantages that may occur in malfunctioning partnerships. Considering the first question, positive prospects for the industry should lead to positive return expectations despite problems that may occur during partnership. In line with literature, I therefore expect to observe positive value creation effects of collaborative efforts.

The second question would be whether certain groups would benefit more than others from cooperation. For example, Kasperk and Drauz (2013) differentiate between different market participants involved: For automotive firms, they see particularly the interest of developing technological competence. Whereas service companies and power suppliers rather focus on developing new mobility and infrastructure concepts. In other industries, however, previous results also show that firm size may play an important, if not the most important role (Campart and Pfister, 2007; Danzon et al., 2005). This is in line with Kasperk and Drauz (2013) who see partnerships as an important financial funding source for small firms in the automotive industry. Thus, the role of small-sized ventures and their financial situation is surely an important aspect when analyzing determinants of value creation.

Another idea would be that the level of intensity and type of partnership influences the value creation potential. The level of cooperation intensity determines the flexibility and stability of the relationship. At one extreme one would have rather lax contractual (horizontal, vertical) agreements which allow maximum flexibility at low relationship intensity, while conversely at the other extreme strategic alliances may offer higher stability and demand a higher commitment at the same time. However, the intensity level of partnerships is empirically difficult to determine, especially from outside. This might be the reason why many studies focus on Joint Ventures which are usually explicitly communicated as such. Considering the type, it is especially R&D activity which tends to create value, since asymmetric information structures and high development costs may be beneficial for cooperations (Campart and Pfister, 2007). Besides technological partnerships, Proff and Proff (2013) see positive effects of cooperation for projects in the automotive industry with high fix and launching costs. This especially applies to all infrastructural projects that come with E-mobility. Concerning the second question, I expect especially positive effects for small-sized ventures, when collaborating with a larger firms. In addition, I would particularly expect positive returns for R&D-driven cooperations. Other influential factor may play a role and will be tested by regression analysis.

A brief outline of this chapter is as follows: Section 4.1 shortly discusses some background information, main motivations and technological issues which lead to collaboration in the area of EV technologies. Section 4.2 presents and discusses the hand-collected international data set on e-mobility cooperations and its creation. In section 4.3 I shortly introduce the methods applied, which is mainly the standard market model event study methodology and multivariate regression analysis for finding the most important determinants. The last two sections, present the results and discuss possible implications.

4.1. Background: Cooperation for E-mobility, main goals and motivation

As in many industries, Small and Medium-sized Enterprises (SMEs) are regularly (but not exclusively) the driving force behind innovation and often possess relevant knowledge but lack of resources and capabilities (Nieto and Santamar'ıa, 2010). For E-mobility, it is especially comparable young Asian firms with advanced battery knowledge, originally gained by other products of the IT sector, which might benefit from the collaboration. For the long-established automotive industry, on the other hand, this leads to the question of which technologies along the supply chain are indispensable and how to integrate external knowledge into own supply chains. By literature review and market observation, onecan identify mainly three areas which motivate collaborative activities for electric vehicles: 1. Technological uncertainty and risks (e.g. Wallentowitz et al., 2008) 2. the mastery of the whole product lifecycle (e.g. Morche et al., 2013) and 3. testing new business models and concepts and the controlling of marketing related risks (e.g. Kasperk and Drauz, 2013). The identified motivations will be briefly outlined as follows:

The access and the need for temporal, horizontal and vertical integration of technologies into the production process can be considered as one of the main motivations, if not the most important motivation, for R&D-cooperation. Besides a general R&D trend and shift from mechanical systems over mechatronics to electronics, literature and the data set suggest that 5 main technological areas can be identified which drive cooperative structures in the automotive industry (Wallentowitz et al., 2010; Maier, 2010; Hanselka and Jöckel, 2010):

- Energy storage (Fuel - Fuel and battery - Battery - fuel cell)
  A major obstacle for e-vehicles can still be identified in the energy storage problem (Maier, 2010). While many alternative storage possibilities have been proposed by engineers, it is still unclear which technology will win the race in the end. All proposed technological solutions still face problems of low energy densities and significantly increase vehicle weights (Maier, 2010).[16] This counterproductive characteristic should be offset by the positive characteristics of the efficient electric power train. An overview of the mostly discussed storage/power train/infrastructure solutions is provided in table 4.1.
- Power Train (Fuel - Hybrid - Pure electric drive)
  The power train block faces an evolution from pure combustion engines via hybrid drives to pure electric drives (Kohler, 2010). While pure electric drive solutions are clearly the most efficient way of energy conversion to kinetic energies, the constraint of pure electric power storage still leads to very restricted driving ranges of approx. 100 miles. Toyota, as the first manufacturer of series-hybrid cars started this development in 1997, and it soon became clear, that the hybrid may close the necessary technology gaps between pure fossil and pure electric driven vehicles. However, power train technologies are very complex and most carmakers must hold ready plans for the whole pallet from combustion engines to pure electric drives.
- Standardization (Connections and specifications)
  A third technological block deals with necessary connections and specifications. While issues of standardization may seem to be a minor problem, they still mark considerable barriers for worldwide producers and customers. Besides AC/DC connectors, many different voltages, capacities and battery

16 Even though it is sometimes argued that conventional combustion engine vehicles do not need to transport the necessary oxygen which in turn would lead to lower energy densities of fossil energy sources. However, the argument does not really concern customers and is relatively weak in the end.

systems are in the market and it takes a lot of time and costs to agree on one widely accepted standard or to set an own standard.

- Infrastructure solutions for energy supply (Gas station - Electric filling station - Battery change)
  In distinction to technological decisions which directly affect the vehicle design, the question of how the energy supply should be organized is still unsolved and also affects business models of power supply companies (and may in turn also affect vehicle design decisions). While most low-range EVs rather require an access to a building's electrical supply network, other solutions may include public charge stations or even the exchange of whole energy storage units at supply stations.
- Energy efficiency and lightweight construction
  One key technology to EV is the mastery of lightweight construction. Most electricity storage concepts with rather low energy densities increase vehicle weights and may therefore be counterproductive for the goal of weight reduction and energy efficiency. Especially construction knowledge based on carbon fibers may be crucial for offsetting the negative battery storage effects.

Another main motivation for cooperation can be spotted in the coverage of new resource demands and the mastery of life cycle processes (Wallentowitz et al., 2008). The issue of rare resources is closely linked with the problem of recycling and reuse of valuable materials in batteries and other components. The problem has not been widely discussed in literature but a few R&D-cooperations explicitly declare the goal of developing recycling methods for E-mobility concepts.

Besides technological issues and the mastery of product lifecycles, testing new business models can be seen as another motivation for collaboration (Kasperk and Drauz, 2013). Especially traditional power producers see that electric vehicles necessitate a completely different infrastructure and hope, if only partly, to replace oil firms and their gas stations. As mentioned above, the three named key motivations also mirror in the data set which is discussed in the following section.

Table 4.1.: Proposed solutions and evolution from combustion engines to fuel cell powered vehicles. Own representation based on VDA (2013).

Combustion engines	**Hybrid**	**Plug-In Hybrid**	**Range Extended Electric Vehicle**	**Battery powered vehicle**	**Fuel cell powered vehicle**
Conventional diesel and gas engines with improvement and saving potential.	Combination of electric engine and conventional engine. The battery is charged by driving and/or braking energy.	In addition to hybrid, plug-in hybrid also allows to load the battery via the power supply system.	When needed, a conventional engine generates necessary energy to extend driving range.	An electric vehicle solely powered by batteries. Loading via the power supply system.	An electric vehicle powered by hydrogen or methane. Energy supply ideally produced by renewable energies.

4.2. Data

The cooperation study is based on a hand-collected data-set. The data set comprises all observed cooperations between listed firms that deal with E-mobility from 1995 to January 2012. Therefore, I initially selected relevant firms of the automotive and electronic device sector. A first hint for a relevant firm list was provided by Bloomberg's new energy finance data base. In addition to database related data, I extracted information out of newspaper articles and by considering firm's related information in LexisNexis. The respective event dates were set to the point in time when the new information of the cooperative event firstly enters the market. In most cases, the relevant event date was determined by the respective firm's press release, in few cases by press articles which informed the market participants earlier. Furthermore, it is required that all involved cooperative firms are listed in the stock market. While this requirement could be relaxed in future studies, it reduces a possible bias toward larger firms and allows to direct comparison of the effect on both partners. Overall, the data set contains 64 hand-picked cooperative events and 114 affected firms. The total return data originates from Thomson Reuters Financial database. Table 4.2 summarizes the cooperative events. In most cases, the cooperative events are between two firms. However, event 38 is between three partners and event 55 between five. In addition to the event date and the affected firm, the list also contains whether the firms explicitly communicated a Joint venture. While it could have been interesting to determine other types of cooperation as well, most firms did not use a standardized wording in their press releases. For example, the term strategic "alliance" does not tell much about the depth of the cooperation and whether supply contracts play a role or not. As market benchmarks, I choose associated Thomson Reuters total return equity indexes for the respective country. The indexes aim at a comprehensive and rule based coverage with high liquidity and market capitalization requirements (ThomsonReuters, 2013). In order to form sensible subsamples, the firms are also separated by size. The intuition behind is that I intend to analyze whether small firms benefit above average from cooperation. As an indicator for size market capitalization is chosen and the market cap of the larger company must at least be three times the market cap of the smaller company. By this, a subsample of small firms, large firms and cooperations between "equals" can be identified. The sole exception of this rule is in table 4.2 event 55 which includes one smaller company (EnBW) but reasonably still represents a cooperation between 'equals' in terms of a large-sized power producer which cooperates with the automotive companies.

Table 4.2.: Hand-collected data set. The table lists event-dates, the firms affected and whether the announced cooperation was a joint venture.

Event no.	Date	Firm name 1	Firm name 2	Joint venture
1	20-05-1996	Toyota Motor	Panasonic	yes
2	03-26-2004	Hitachi	Shin Kobe Elec.	yes
3	06-30-2004	GP Batteries Intl.	Sanyo Electric	yes
4	04-13-2007	NEC	Nissan Motor	yes
5	05-08-2007	Mitsubishi Motors	GS Yuasa	yes
6	07-23-2007	Azure Dynamics	Ford Motor	
7	05-27-2008	Toyota Motor	Panasonic	yes
8	05-28-2008	Sanyo Electric	Volkswagen	
9	06-16-2008	Peugeot	Mitsubishi Motors	
10	10-14-2008	Electrovaya	Tata Motors	
11	12-17-2008	Honda Motor	GS Yuasa	yes
12	01-12-2009	Magna Intl.	Ford Motor	
13	01-13-2009	Azure Dynamics	Johnson Controls	
14	02-02-2009	Itochu	Ener1	
15	02-12-2009	Toshiba	Volkswagen	

Event no.	Date	Firm name 1	Firm name 2	Joint venture
16	04-02-2009	SGL Carbon	Freni Brembo	yes
17	05-25-2009	BYD	Volkswagen	
18	06-23-2009	Electrovaya	Exxon Mobil	
19	07-13-2009	Volvo	Ener1	
20	07-31-2009	BMW	E.ON	
21	08-31-2009	Hyundai	LG Chem	yes
22	09-03-2009	Borgwarner	UQM Techs.	
23	09-15-2009	RWE	Renault	
24	09-25-2009	Volkswagen	Johnson Controls	yes
25	10-29-2009	BMW	SGL Carbon	yes
26	10-30-2009	Azure Dynamics	Ford Motor	
27	11-02-2009	Peugeot	Sanyo Electric	
28	01-04-2010	Tianneng Power Intl.	Kandi Technologies	
29	01-11-2010	Balqon	Ford Motor	
30	03-01-2010	Daimler	BYD	yes
31	03-11-2010	Borgwarner	Azure Dynamics	
32	04-26-2010	Volvo	LG Chem	
33	04-28-2010	Daimler	Toray Inds.	yes
34	05-13-2010	Suzuki Motor	Sanyo Electric	yes
35	06-18-2010	Daimler	ENBW	
36	07-02-2010	Toshiba	Mitsubishi Motors	
37	07-13-2010	Ford Motor	LG Chem	
38	09-15-2010	Daimler Renault	Nissan Motor	
39	09-16-2010	BYD	RWE	
40	09-17-2010	Magna Intl.	GS Yuasa	yes
41	09-29-2010	Mitsubishi Motors	LG Chem	
42	10-29-2010	Toyota Motor	Mazda Motor	
43	11-23-2010	Itochu	Duke Energy	

44	01-10-2010	ABB	Ecotality	
45	01-24-2011	Daimler	Toray Inds.	yes
46	02-01-2011	BMW	Peugeot	
47	02-09-2011	SK Innovation	Daimler	
48	04-08-2011	Peugeot	GS Yuasa	
49	07-14-2011	Daimler	Ecotality	
50	08-23-2011	Ford Motor	Toyota Motor	
51	08-31-2011	Siemens	Volvo	
52	11-29-2011	Toyota Motor	Panasonic	
53	01-10-2012	SK Innovation	Continental	yes
54	01-11-2012	A123 Systems	BMW	
55	01-23-2012	BMW ENBW Siemens	Daimler RWE	yes

4.3. Method

This study aims to attain two methodic goals. The first goal is to measure the effects of EV related corporate cooperations on the valuation of stock equity. This carried out by standard market model event study methodology (Brown and Warner, 1985; MacKinlay, 1997; Fuller et al., 2002). The second goal is to apply a multivariate Ordinary Least Squares (OLS) regression in order to find the most important effect determinants. The following section explains both approaches more in detail.

The goal of an event study is to detect non-predictable or so-called abnormal returns surrounding the event date. The Cumulative Abnormal Returns (CAR) in the market model is defined as:

$$CAR_{i,t_1,t_2} = \sum_{t_1=t_0-wi_1}^{t_2=t_0+wi_2} Re_{i,t} - \widehat{\alpha} - \widehat{\beta} Re_{Index,t} \tag{4.1}$$

,where $Re_{i,t}$ is the return of the delineated firm at time t, $\hat{\alpha}$ and $\hat{\beta}$ are the regression estimates from OLS regression and $Re_{Index,t}$ is the return of the respective market index. Time t_0 marks the event date, wi_1 and wi_2 define the event window and may be equal for symmetric time frames. For parameter estimation, I apply White regression instead of OLS in order to address possible issues of heteroskedacity (White, 1980; Davidson and MacKinnon, 1993). The estimation period is 180 days and ends 30 days before the event date t_0 to make sure the event can not influence the estimation results. The Cumulated Average Abnormal Return (CAAR) for the whole sample is defined as the average of all N observations:

$$CAAR_{t_1,t_2} = \frac{\sum_1^N CAR_{i,t_1,t_2}}{N} \tag{4.2}$$

Besides ordinary t-test, Boehmer test and Wilcoxon signed rang test (non-parametric) is carried out for testing the hypothesis that CAAR is significantly different from zero (Boehmer et al., 1991; Wilcoxon, 1945; Siegel, 1956).

For explaining the determinants of cumulated abnormal returns we apply simple multivariate OLS regression of the form

$$CAR_{i,t_1,t_2} = \hat{\gamma} + \hat{\delta} X + \xi \tag{4.3}$$

and test several quantitative and one qualitative influential parameter X. Residual analysis is performed to confirm proper and linear model specification.

4.4. Results

The results section is divided into two subsections. The first subsection presents results from event study calculations, the second elucidates the results from multiple regression and possible explanations of the abnormal returns.

4.4.1. Event study results

The results for the whole data set are summarized in table 4.3. With the help of n=114 observations, I find cumulative abnormal returns between 2% and 2.7% sourrounding the corresponding event dates. Some of the effect at 0,1% significance level can be attributed to the event interval [–2; 0] which indicates that the information may leak earlier into the market. Besides p-values, the other test statistics also indicate highly significant results at 1% significance level. However, rather low median values indicate a skewed distribution of the returns. The median return for the event date [0; 0] is at 0,3% only.

Table 4.3.: Results I: The table summarizes the results for the complete data set with n=114 observations. Window *indicates the event window surrounding the event date.* Mean CAR *is the cumulated average abnormal return (CAAR) for all observations.* Median CAR *calculates the median of the abnormal returns.* T-stat *is the standard t-student test static.* Boehmer-z-score *is the calculation of Boehmer test according to Boehmer et al. (1991).* Wilcoxon z-score *is a non-parametric rank test (Wilcoxon, 1945). The* p-values *assign the probabilities to the respective test statics.*

Window		Mean CAR (CAAR)	Median CAR	t-stat	p-val	Boehmer z-score	p-val	Wilcoxon z-score	p-val	Nobs
0	0	0.020	0.003	3.405	<0.001***	3.020	0.003**	-3.052	0.002**	114
-1	1	0.024	0.009	3.179	<0.001***	2.913	0.004**	-3.131	0.002**	114
-2	2	0.022	0.007	2.711	0.007**	2.682	0.008**	-2.371	0.018**	114
-5	5	0.012	0.009	1.046	0.296	1.127	0.262	-0.728	0.467	114
-10	10	0.013	0.009	0.760	0.447	0.826	0.411	-0.259	0.796	114
-20	20	-0.018	-0.007	-0.628	0.530	-0.771	0.442	-0.810	0.418	114
-10	0	0.029	0.009	1.630	0.103	1.654	0.101	-1.158	0.247	114
-5	0	0.025	0.005	2.942	0.003**	2.863	0.005**	-2.391	0.017*	114
-2	0	0.027	0.008	3.342	<0.001***	3.129	0.002**	-3.632	<0.001***	114
-1	0	0.025	0.011	3.495	<0.001***	3.006	0.003**	-4.129	<0.001***	114
0	1	0.019	0.004	2.928	0.003**	2.885	0.005**	-2.405	0.016*	114
0	2	0.015	0.007	2.361	0.018*	2.259	0.026*	-2.167	0.030*	114
0	5	0.007	0.003	0.622	0.534	0.787	0.433	-0.550	0.582	114
0	10	0.005	0.007	0.621	0.534	0.515	0.607	-0.728	0.467	114

Significant at ***≤ 0.1%-, **≤ 1%- and *≤ 5%-significance-level.

Table 4.4.: Results II: Subsample of 31 small-sized firms. The firm's market cap is at least three times smaller compared to their larger partners.

Window		Mean CAR (CAAR)	Median CAR	t-stat	p-val	Boehmer z-score	p-val	Wilcoxon z-score	p-val	Nobs
0	0	0.070	0.029	3.852	<0.001***	3.249	0.003**	-4.174	<0.001***	31
-1	1	0.087	0.045	3.954	<0.001***	3.400	0.002**	-4.154	<0.001***	31
-2	2	0.082	0.043	3.736	<0.001***	3.308	0.002**	-3.645	<0.001***	31
-5	5	0.067	0.057	2.591	0.010**	2.362	0.025*	-2.234	0.025*	31
-10	10	0.107	0.059	2.894	0.004**	2.298	0.029*	-2.175	0.030*	31
-20	20	0.026	0.037	0.700	0.484	0.369	0.715	-0.607	0.544	31
-10	0	0.121	0.057	2.616	0.009**	2.153	0.040*	-2.900	0.004**	31
-5	0	0.083	0.050	3.774	<0.001***	3.306	0.002**	-3.547	<0.001***	31
-2	0	0.083	0.034	3.387	0.001***	2.978	0.006**	-4.096	<0.001***	31
-1	0	0.083	0.041	3.579	<0.001***	3.024	0.005**	-4.664	<0.001***	31
0	1	0.074	0.036	4.200	<0.001***	3.764	0.001***	-3.723	<0.001***	31
0	2	0.069	0.059	3.995	<0.001***	3.511	0.001***	-3.488	<0.001***	31
0	5	0.053	0.024	2.481	0.013*	2.269	0.031*	-2.646	0.008**	31
0	10	0.055	0.045	2.950	0.003**	2.622	0.014*	-2.214	0.027*	31

Significant at ***≤ 0.1%-, **≤ 1%- and *≤ 5%-significance-level.

Table 4.5.: Results III: Subsample of 31 large-sized firms. The firm's market cap is at least three times smaller compared to their larger partners.

Window		Mean CAR (CAAR)	Median CAR	t-stat	p-val	Boehmer z-score	p-val	Wilcoxon z-score	p-val	Nobs
0	0	0	-0.002	0.027	0.979	0.090	0.929	-0.137	0.891	31
-1	1	-0.004	-0.008	-0.603	0.546	-0.523	0.605	-0.059	0.953	31
-2	2	-0.009	-0.009	-1.168	0.243	-1.195	0.241	-1.019	0.308	31
-5	5	-0.012	-0.024	-0.883	0.377	-0.832	0.412	-0.764	0.445	31
-10	10	-0.044	-0.048	-2.115	0.034*	-2.199	0.036*	-2.234	0.025*	31
-20	20	-0.065	-0.070	-2.065	0.039*	-2.160	0.039*	-2.312	0.021*	31
-10	0	-0.017	-0.025	-1.211	0.226	-1.386	0.176	-1.293	0.196	31
-5	0	0.001	-0.003	0.016	0.987	0.062	0.951	-0.137	0.891	31
-2	0	0.001	-0.008	0.002	0.999	0.207	0.837	-0.176	0.860	31
-1	0	-0.002	-0.005	-0.331	0.741	-0.316	0.754	-0.255	0.799	31
0	1	-0.002	0.002	-0.395	0.693	-0.275	0.785	-0.039	0.969	31
0	2	-0.009	-0.001	-1.450	0.147	-1.419	0.166	-0.882	0.378	31
0	5	-0.013	-0.020	-1.169	0.242	-1.134	0.266	-1.019	0.308	31
0	10	-0.027	-0.015	-1.709	0.087	-1.576	0.126	-1.352	0.176	31

Significant at ***≤ 0.1%-, **≤ 1%- and *≤ 5%-significance-level.

Table 4.6.: Results IV: Subsample for firms without clear size differences.

Window		Mean CAR (CAAR)	Median CAR	t-stat	p-val	Boehmer z-score	p-val	Wilcoxon z-score	p-val	Nobs
0	0	0.002	-0.001	1.045	0.296	0.925	0.359	-0.656	0.512	52
-1	1	0.004	0.003	1.302	0.193	0.966	0.339	-1.157	0.247	52
-2	2	0.005	0.006	1.274	0.203	1.027	0.309	-1.020	0.308	52
-5	5	-0.006	0.003	-0.296	0.767	-0.714	0.479	-0.446	0.655	52
-10	10	-0.007	0.003	-0.197	0.844	-0.511	0.612	-0.082	0.935	52
-20	20	-0.013	0.004	-0.234	0.815	-0.652	0.517	-0.137	0.891	52
-10	0	0.001	0.002	0.041	0.967	0.119	0.906	-0.273	0.785	52
-5	0	0.006	0.003	0.983	0.326	0.849	0.400	-0.637	0.524	52
-2	0	0.009	0.008	1.864	0.062	2.096	0.041*	-2.304	0.021*	52
-1	0	0.008	0.008	2.125	0.034*	1.920	0.060	-2.022	0.043*	52
0	1	-0.001	0	-0.034	0.973	-0.319	0.751	-0.137	0.891	52
0	2	-0.002	0.003	0.145	0.884	-0.396	0.694	-0.765	0.444	52
0	5	-0.009	0	-0.906	0.365	-1.298	0.200	-0.801	0.423	52
0	10	-0.006	0.005	-0.042	0.966	-0.609	0.545	-0.036	0.971	52

Significant at ***≤ 0.1%-, **≤ 1%- and *≤ 5%-significance-level.

The observation leads to the question of whether other effects may help to find a better explanation of the results. As suggested by other authors, I therefore test for size-effects and construct corresponding subsamples. An easy way to do this, is by demanding that the larger partner should have at least a three times higher market capitalization when the event was announced. By this approach, I identify 62 firms with notable size difference and another 52 firms which have a cooperation "between equals." Table 4.4 shows the results for the 31 smaller-sized partner firms. The results show great cumulated average abnormal returns at the event date of around 7% with very high significance levels. Medians at 3% point out positive overall effects but also indicate distribution skewness. In contrast, table 4.5 does not confirm the same effect for the larger sized partners. In contrary I find rather slight negative returns and cannot find statistical significant results that would confirm returns different from zero at the event date. Almost the same applies for cooperations among "equals" in table 4.6. For different event windows, positive and negative returns can be observed but the results are of no statistical significance. Even though event windows [−2, 0] and [−1; 0] indicate slightly positive abnormal returns with significant results, this result should not be overstated. Overall, it seems that partnerships between equals do not create short-term wealth effects.

4.4.2. Determinants

Besides the calculation of abnormal returns, I also see the necessity to identify key determinants which may help to explain the identified effects. It may be stated that the key drivers of partnerships are less known than determinants of other capital market activity, such as M&A transactions for instance. As a key outcome of the subsection above, a strong impact of the firm's size which is in line with the results of Das et al. (1998) can be observed. I reconsider this result by introducing a variable for the market capitalization. With the help of residual plots, however, I also observe that the effect is disproportionately strong for small sized firms. Therefore, the logarithmized market capitalization as a first explanation for the abnormal returns in Model 1 of table 4.7 is considered. By this, a strong negative influence of logarithmized market capitalization with an estimate of $\hat{\beta} = -0.02$ at significant white adjusted p-values can be verified.

In literature, another determinant variable widely discussed is the type of partnership (Koh and Venkatraman, 1991; Das et al., 1998; Campart and Pfister, 2007). While Koh and Venkatraman (1991) focus on Joint Ventures, Das et al. (1998); Campart and Pfister (2007) also take technological and production/

marketing alliances into consideration. However, the determination of the cooperative type may be sometimes ambiguous, as for example production alliances may also include R&D related elements. Therefore, I consider Joint ventures only, as they are usually communicated clearly in press releases and may be an obligation for commencing operations in some countries. I furthermore assume, that technological alliances should be driven by R&D efforts and may therefore be identified by R&D-expenditures. Variable "R&D to NetSales" adjusts for possible firm size effects and may determine technological focused firms, especially in the advanced battery sector. In addition, I argue that by considering the E-mobility sector a selection of more technology driven partnerships is more likely. In Model 2, however, I find neither variable useful for a better explanation of the overall model. While R&D-expenditures tend to have slightly positive effects, Joint ventures rather exhibit the opposite.

Another widespread argument is that investment decisions which come in line with partnerships may be strongly influenced by positive investment prospects (Brooke and Oliver, 2005). While (Brooke and Oliver, 2005) suggest market-to-book ratios, (Das et al., 1998) rather focus on the profitability of the firm by considering ROI's. I follow both approaches by using market-to-book ratios as well as Earnings Before Interests, Taxes, Depreciation and Amortization (EBITDA) margins for profitability. Model 3 adds the two investment oriented variables which show positive coefficient effects, but without any significance.

Summing up the results of the multiple regression, it can be stated that size effects have the strongest and most significant impact on my data sample. Small size firms do disproportionally benefit from cooperations in the industry. The result stay somewhat in contrast to findings of previous studies in other industries which might suggest other determinants may also play an important role. This raises the question of what makes the industry unique and different from other technology driven business areas, such as the pharmaceutical industry or information technology industry. To my mind, however, the main difference may probably determined by the uncertainty of whether electric vehicles will make it in the future. The latest developments in shale gas, improvements of traditional combustion engines and the non deniable disadvantages of electric energy storage might still be predominant. Seeing the results in that light, it is clear that the best utilization of technology for small firms lies in cooperation while big players might give themselves a "green" image and reserve all technological options.

Table 4.7.: Results from multiple regression analysis. $\hat{\beta}$ is the regression estimate, t-stat is the t-statistic, calculated with white-adjusted standard errors and VIF the respective variance inflation factors. MarketCap is the market capitalization, R&D to Rev. the R&D expenditures in relation to revenues and EBITDA to Rev. the EBITDA margin. JV-Dummy is a dummy for Joint ventures announcements and Market to book the respective ratio.

	Model 1			Model 2			Model 3		
Variable	$\hat{\beta}$	**t-stat**	**VIF**	$\hat{\beta}$	**t-stat**	**VIF**	$\hat{\beta}$	**t-stat**	**VIF**
Constant	0.362	3.02***	-	0.459	5.81***	-	0.464	5.74***	-
log(MarketCap)	-0.02	-2.98***	1	-0.019	-5.60***	1.34	-0.019	-5.52***	1.34
R&D to Rev.				>-0.1%	-1.37	1.63	>-0.1%	-1.13	1.83
EBITDA to Rev.				0.007	0.93	1.87	0.008	1.00	1.90
JV-Dummy				0.007	1.02	1.03	0.008	1.01	1.04
Market to book							<0.1%	0.42	1.14
N obs		114			95			95	
R^2		0.2337			0.2802			0.2881	
R^2 adjusted		0.2345			0.2482			0.2481	
F-stat		34.15			8.75			7.20	
P-value (F-stat)		<0.1%***			<0.1%***			<0.1%***	

Significant at ***≤ 1%-, **≤ 5%- and *≤ 10%-significance-level.

4.5. Discussion of chapter 4

This chapter evaluated effects of cooperation in the automotive industry, in particular in the area of E-mobility. On the one hand, the observation of many cooperative events instead of other capital market activity suggest a rather cautious estimation of carmakers concerning the future technological prospects. On the other hand, the high intensity and sheer number of cooperations might indicate that the industry is at an early emerging stage with a lot of uncertainty addressed by the means of partnerships. Since markets evaluate future prospects of ventures very early, event study methodology may be an appropriate method to evaluate the assessment of the market for that technology. With the help of the event study methodology, I found very positive overall short-term evaluation effects, but not for all partners in the same manner. It can be stated, that it is especially the small firms which strongly benefit from cooperation, large-sized OEMs or suppliers sometimes suffer even from negative effects. To interpret this, one may assume that small technological ventures are extremely risky and the announcement of a cooperation with a much bigger-sized partner rouses speculation of possible commercialization and positive financial boost. For the bigger-sized partner, however, the cooperation usually comes in line with new financial obligations and uncertain advantages for series production. The same applies for cooperations among "equals" which might not be sufficiently significant for value effects on a stock market scale. Astonishingly, the multiple regression analysis could not detect additional determinants besides size-effects. While studies from other industry suggested that the type, stage or investment opportunity of a cooperation might be a value driver, I cannot confirm any of those effects. Instead, I even affirm the previous result that size in terms of market capitalization is the main value determinant. This result, however, may also be interpreted with a more negative storyline. It seems that small ventures provide the required innovative R&D and therefore benefit strongly from collaboration, while the market does not believe that E-mobility will be a huge success for established firms. Without going too much into risk, it seems that they prefer to bear the costs of collaboration.

5. Conclusive remarks and outlook

> "Electric cars are not going to take the market by storm, but it's going to be a gradual improvement."
>
> —Carlos Ghosn

In this final chapter, I briefly recapitulate the main results of each chapter in order to derive a synthesis of this dissertation and to identify some gaps for further research.

Chapter 2 contains the first and probably most complete bibliometric assessment of German environmental economics research. On the one hand, the study focuses on highlighting publications with the most impact, on the other, co-citation methodology has been used to identify the most relevant literature in the context of clean tech research. Remarkably, the results of both areas do not necessarily match, although they base on the same data set. Cleantech related research has not created the most impact in the past and can rather be considered as a side issue of environmental research. However, some monographic works and many books in the area of E-mobility and Cleantech applications could indicate that researchers might find it difficult to publish in the delineated journals.

In the scientific discourse, it is important to keep in mind that the special position of sustainable technologies can be explained by their potential to reduce environmental impact and usually are attributed to reduce external effects. As stated in the introduction but also mentioned by Heymann et al. (2013), it will be crucial for the idea of E-mobility and the entire industry to prove that the climate footprint and overall environmental impact of EVs will be superior in comparison to conventional vehicles. Otherwise, it will be difficult for economists and policymakers to justify governmental support for E-mobility and to establish further market rules for conventional drives. It is certainly conceivable, that today's critical view on Germany's Renewable Energy Act could be transferable to governmental support of E-mobility concepts in the near future if the idea turns out to be too expensive and without sufficient environmental or social benefits.

In Chapter 3, concrete sourcing risks of EVs and Cleantech applications are analyzed by the example of REEs. The chapter makes clear that a switch of technologies usually comes in line with new supply chains and commodity needs and therefore create new kinds of dependencies. The analysis proves that the REE markets are incomplete; producers that rely on those materials cannot protect themselves by hedging against possible price risks. The reason is that potential financial providers will not be able to quote a reasonable price which would reflect an adequate premium for the price risks. The degree of market incompleteness and the associated risk is substantial, as the prices of many elements exploded within a view weeks to more than tenfold levels. The case of REEs dramatically shows that sourcing risks require early strategic management decisions of firms. Avoidance and substitution is probably the most important strategy, but there are still many applications in the industry, such as super small and powerful electric motors, which cannot be produced without the use of REEs. Thus, clever product design and recycling will play a key role in the future. On the other hand, it is the government which will be challenged in the future. The creation of fundamental framework conditions and securing raw material supply lies in the strategic interest of all states (in the western world). From EU's perspective, politicians should aim at the abolition of trade restrictions and customs barriers, negotiate new trade agreements or think about strategic stockpiles. For the case of REEs, it is also the interest of the supply side to reduce price risks. First rumors in some articles suggest that China is interested in building up a commodity stock exchange, a project that should be supported by European Union member states. From my personal point of view, the chapter also reveals a strong demand for further financial research. The determination of commodity scarcity is nowadays subject to geological surveys or economic studies which argue on the basis of market concentration (e. g. monopolies) or aggregate demand and supply studies. The problem with those studies is that they do not tell much about concrete pricing risks of market participants. Contrary to this, the concept of market completeness directly determines the associated price risks. Therefore, it could be very beneficial to create a "commodity map" based on the degree of market incompleteness, not only for REEs, but also for other urgently needed commodities, such as helium. Hedgeability, in terms of market completeness, could therefore become a key indicator for possible sourcing risks and a valuable source for management decisions.

Chapter 4 deals with the capital market perception of E-mobility concepts. The results allow various interpretations for the emerging industry. First of all, capital market activity can be seen as an health indicator for the condition of an industry and for the possibilities of refinancing. For the industry with many young, innovative ventures, it is essential to gain capital market access for funding R&D expenditures or growth opportunities. While IPO, SEO or M&A activity is still on a relatively low level, cooperations between large-sized and small-sized firms can be observed regularly in the automotive industry, especially in the area of E-mobility. The absence of strong capital market activity can be interpreted as a first sign of high market uncertainty, it seems that OEM's prefer cooperations over concrete acquisitions. The observed cooperative structures raise the question of whether cooperations have similar positive funding effects for the smaller partners. The results suggest that this is indeed the case, smaller partners do regularly benefit from cooperations with established firms. However, from outside's position, it is difficult to determine whether the positive evaluation of cooperations comes from direct funding of R&D-activities or from possible future sales prospects of the smaller ventures.

The picture changes somewhat if the large-sized partners, mostly OEMs, are considered. Stock market returns of large-sized firms do not benefit from cooperations, in contrary, if there is any effect, it is usually a slightly negative abnormal return. While funding may not be the key motivation for E-mobility cooperations of large-sized firms, the findings do not suggest that stock markets believe in a prompt success of the E-mobility concepts. However, it should also be stated that E-mobility cooperations usually make up only a very small fraction of OEM's operations and the results should therefore not be overrated. A very similar picture applies for cooperations among equals which do not show significant reactions of capital markets. The results of Chapter 4 suggest that E-mobility is by no means an automatic sell. Small-sized firms could therefore easily come into distressed situations with insufficient funding opportunities, especially if the larger-sized partners decide to dissolve the partnership. Under uncertainty, cooperations seem to be the management method of choice for gaining the necessary knowledge without going to much into the risk of market failure. For future research, it makes sense to reevaluate the capital market activity of the industry as it matures. In addition, only little is known about long-term financial effects of cooperations, especially for small-sized firms.

To sum up, the results of this dissertation paint a rather gloomy picture of the emerging E-mobility industry. Somewhat negative experiences from related technologies, new uprising sourcing risks and a not-so-enthusiastic capital market do not suggest a fast transition toward E-mobility in the short and medium run. However, this consideration excludes the general technological evolution that is going on in the industry for decades: the transition from mechanical to more efficient and controllable electrical systems. In my opinion, it is therefore very likely that the incremental, evolutionary process toward electric mobility will go on.

A. Appendix of Chapter 1

A.1. Matlab script for basic bibliometric evaluation

```
%%% Main bibliometrics script with function calls %%%
% Calculation of co-citation and coupling matrix
% Calculation of similarity measures and Gmuer (2003)/Schaeffer et al. (2006)
% Christian Babl (2014)

%% Read
Thres=0.199999; %Adapt=Threshold levels [0,1]
[A,txt]=xlsread('Vollerhebung.xlsx', 'XYZ', 'XYZ');...
%read (adapt), matrix A=adjacency matrix
A=A(1:end,5:end);
A(isnan(A))=0;

%% Calculation of co-citation and coupling
B=A'*A; %Co-citation matrix
l=sum(A);% Counts citation frequency (cited)
lc=sum(A');% Counts citation frequency (cites)

%%% Code for couplings (alternatively)
%B=A*A'; %Alternative calculation of couplings

Z=(size(B)); % Calculation of matrix size B

%%% Precalculations
Q=l'*ones(1,Z(1));% Multiplied by l
R=ones(Z(1),1)*l;
L=Q+R; % Sum: Cit(i)+Cit(j)
Lmin=min(Q,R); % Min(Cit(i),Cit(j))

%%% Euclidean distance
Salton_and_Euclid=B./(l'*l).^(1/2);%B equals scalar product
Salton_and_Euclid(isnan(Salton_and_Euclid))=0;

%%% Calculation of Cocit according to Gmuer (2003)/Schaeffer et al. (2006)
```

```
Gmuer=B.^2./(Lmin.*L./2); %Calculation of Cocit-Score
Gmuer(isnan(Gmuer))=0;

%%% Calculation Jaccard-Index
Jaccard=B./(L-B); %Cocup devided by Cit(i)+Cit(j)-Cocup,
  Alternatively: Jaccard=1-squareform(pdist(A','Jaccard'))
Jaccard(isnan(Jaccard))=0;

%%% Calculation Pearson-correlation
Pearson=corrcoef(B);
Pearson(isnan(Pearson))=0;

%% Apply threshold to Gmuer matrix
[Gmuer]=threshold(Thres,Gmuer);

%%% Get clusters
o=1; %Cluster number
[results.cluster.cluster_vektors,~,~,results.cluster.
  amount_of_clusters]=clusterizer(Gmuer);

[~ ,c ,~]=whatcluster(results.cluster.cluster_vektors,txt,o);

%%% Data adjustment
a=find(c<3); %Adapt max. allowed clustersize
Q=sum(results.cluster.cluster_vektors(:,a),2);
Q=logical(Q);
Gmuer(Q,:)=[]; %Delete empty columns and rows
Gmuer(:,Q)=[];
txt(Q,:)=[]; %Delete related text data

%%% Results for each cluster
[results.cluster.cluster_vektors,results.cluster.b,results.
  cluster.all_clusters,results.cluster.amount_of_clusters]=
  clusterizer(Gmuer);
[results.cluster.group_overview,results.cluster.c,results.
  cluster.group]=whatcluster(results.cluster.
  cluster_vektors,txt,o);
```

```
% threshold.m - Applies threshold to matrix
% Christian Babl(2014)
function [B]=threshold(thres,A)

Z=thres<A;
B=A.*Z;
```

```
% clusterizer.m - Function for finding connected clusters.
% Partly utilizes code by Edward Scheinerman, circa 2006,
  isconnected.m, source: http://www.ams.jhu.edu/~ers/matgraph/
% And improvements by Gergana Bounova (2013)/M-file:
  isConnected.m, source: https://github.com/aeolianine/
  octave-networks-toolbox/blob/master/isConnected.m

function [b,u,v,n]=clusterizer(A)

k = length(A);
x = zeros(k,1);

for n=1:k
  m=1:k;
x(n)=1;

while 1
  y = x;
  x = A*x + x;
  x = x>0;
  if x==y
    break
  end
end

q(m,n)=x;
x = zeros(k,1);
end
```

```

[b,u,v] = unique(q(:,1:size(q)), 'rows');
n=size(u,1);
b=b';
```

```
% whatcluster.m - Assignment of apt metadata text to cluster
  groups.
% Christian Babl (2014)
function [txt3,c,d]=whatcluster(a,txt,N)
txt3=txt;
txt3(:,:)=[];
for n=1:size(a,2);
    txt2=txt;
    c= a(:,n)==0;
    txt2(c,:)=[];
    txt3=[txt3;txt2];
end

txt3;
c=sum(a,1);

%%% Cluster number
d=txt3((sum(c(1:(N-1)))+1):sum(c(1:N)),:);

end
```

B. Appendices of Chapter 2

B.1. Rare earth time series

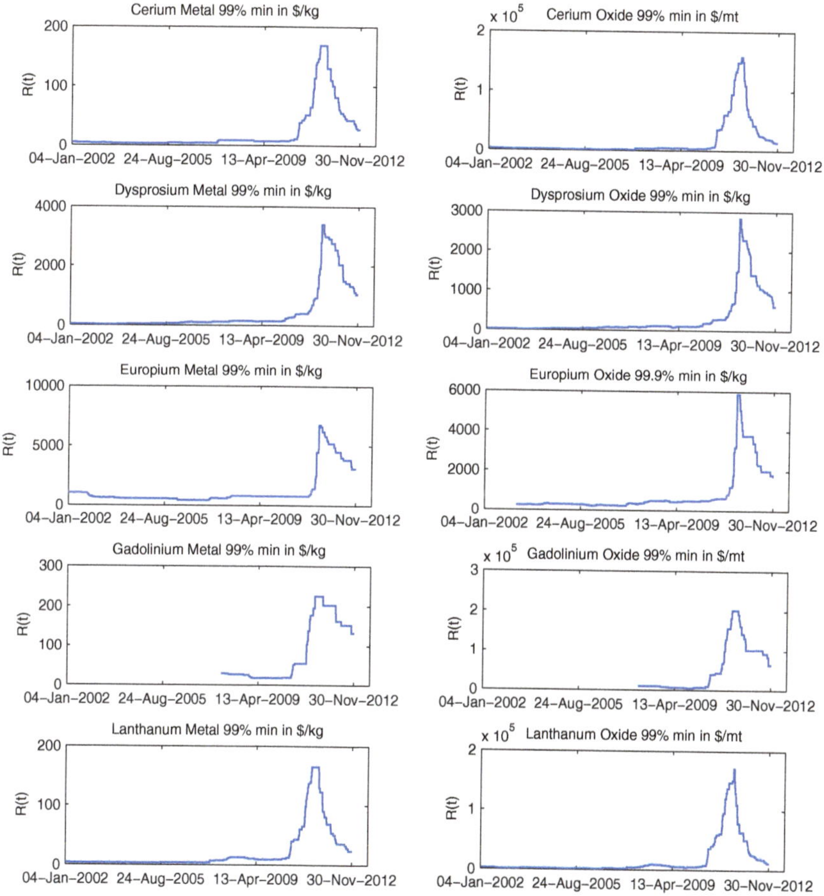

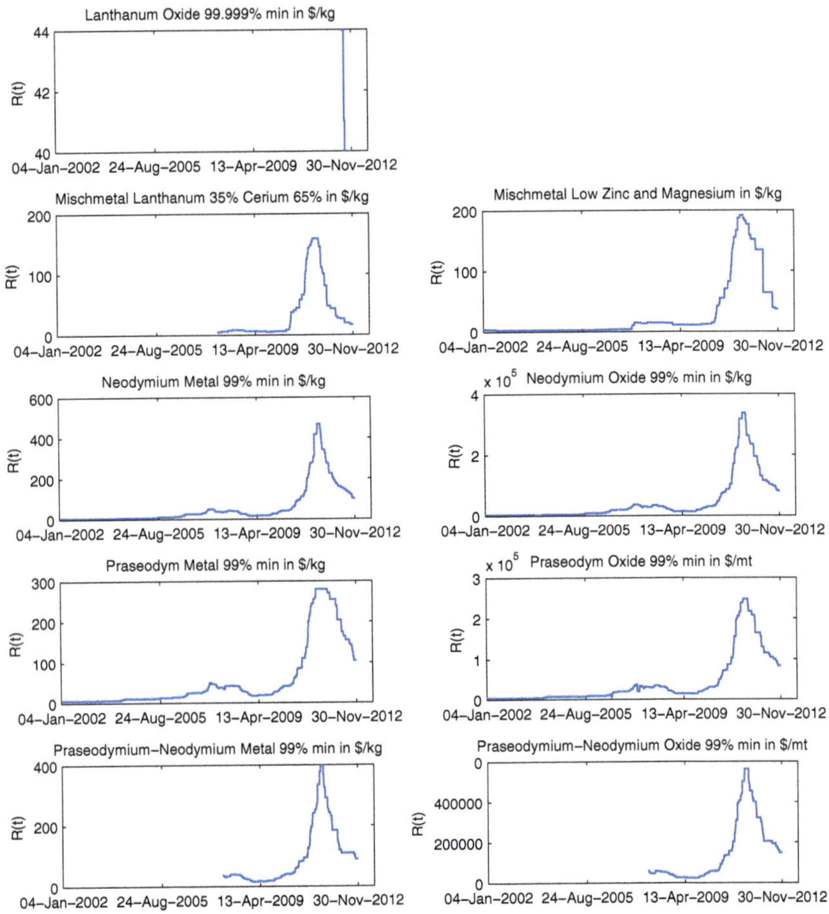

Lanthanum Oxide 99.999% min in $/kg
R(t)
44
42
40
04–Jan–2002 24–Aug–2005 13–Apr–2009 30–Nov–2012
Mischmetal Lanthanum 35% Cerium 65% in $/kg
R(t)
200
100
0
04–Jan–2002 24–Aug–2005 13–Apr–2009 30–Nov–2012
Mischmetal Low Zinc and Magnesium in $/kg
R(t)
200
100
0
04–Jan–2002 24–Aug–2005 13–Apr–2009 30–Nov–2012
Neodymium Metal 99% min in $/kg
R(t)
600
400
200
0
04–Jan–2002 24–Aug–2005 13–Apr–2009 30–Nov–2012
x 10^5 Neodymium Oxide 99% min in $/kg
R(t)
4
2
0
04–Jan–2002 24–Aug–2005 13–Apr–2009 30–Nov–2012
Praseodym Metal 99% min in $/kg
R(t)
300
200
100
0
04–Jan–2002 24–Aug–2005 13–Apr–2009 30–Nov–2012
x 10^5 Praseodym Oxide 99% min in $/mt
R(t)
3
2
1
0
04–Jan–2002 24–Aug–2005 13–Apr–2009 30–Nov–2012
Praseodymium–Neodymium Metal 99% min in $/kg
R(t)
400
200
0
04–Jan–2002 24–Aug–2005 13–Apr–2009 30–Nov–2012
Praseodymium–Neodymium Oxide 99% min in $/mt
R(t)
0
400000
200000
0
04–Jan–2002 24–Aug–2005 13–Apr–2009 30–Nov–2012

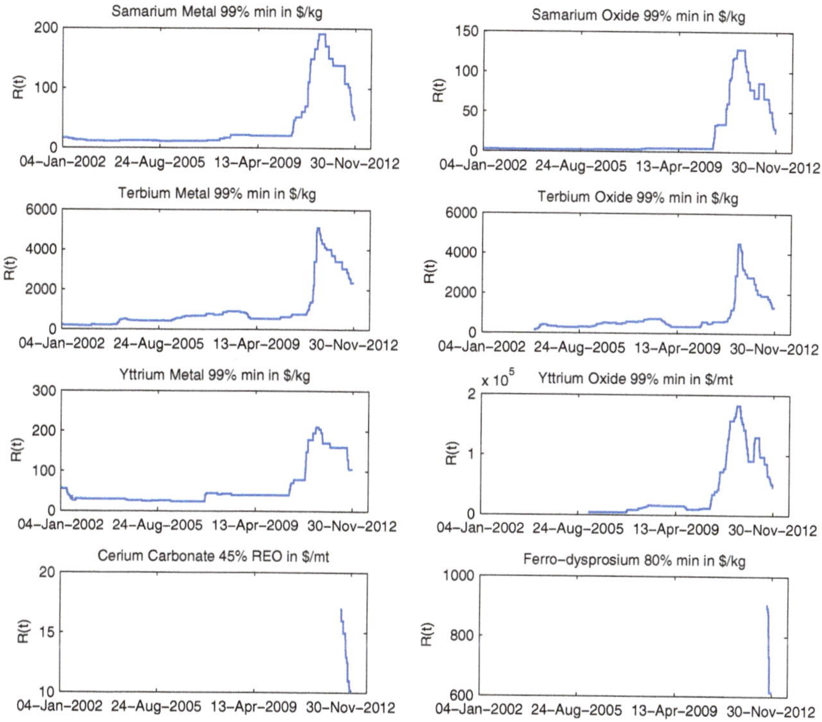
Samarium Metal 99% min in $/kg
R(t)
200
100
0
04-Jan-2002 24-Aug-2005 13-Apr-2009 30-Nov-2012
Samarium Oxide 99% min in $/kg
R(t)
150
100
50
0
04-Jan-2002 24-Aug-2005 13-Apr-2009 30-Nov-2012
Terbium Metal 99% min in $/kg
R(t)
6000
4000
2000
0
04-Jan-2002 24-Aug-2005 13-Apr-2009 30-Nov-2012
Terbium Oxide 99% min in $/kg
R(t)
6000
4000
2000
0
04-Jan-2002 24-Aug-2005 13-Apr-2009 30-Nov-2012
Yttrium Metal 99% min in $/kg
R(t)
300
200
100
0
04-Jan-2002 24-Aug-2005 13-Apr-2009 30-Nov-2012
x 10^5
Yttrium Oxide 99% min in $/mt
R(t)
2
1
0
04-Jan-2002 24-Aug-2005 13-Apr-2009 30-Nov-2012
Cerium Carbonate 45% REO in $/mt
R(t)
20
15
10
04-Jan-2002 24-Aug-2005 13-Apr-2009 30-Nov-2012
Ferro-dysprosium 80% min in $/kg
R(t)
1000
800
600
04-Jan-2002 24-Aug-2005 13-Apr-2009 30-Nov-2012

B.2. Inhomogeneous intensity rates: Lanthanum example

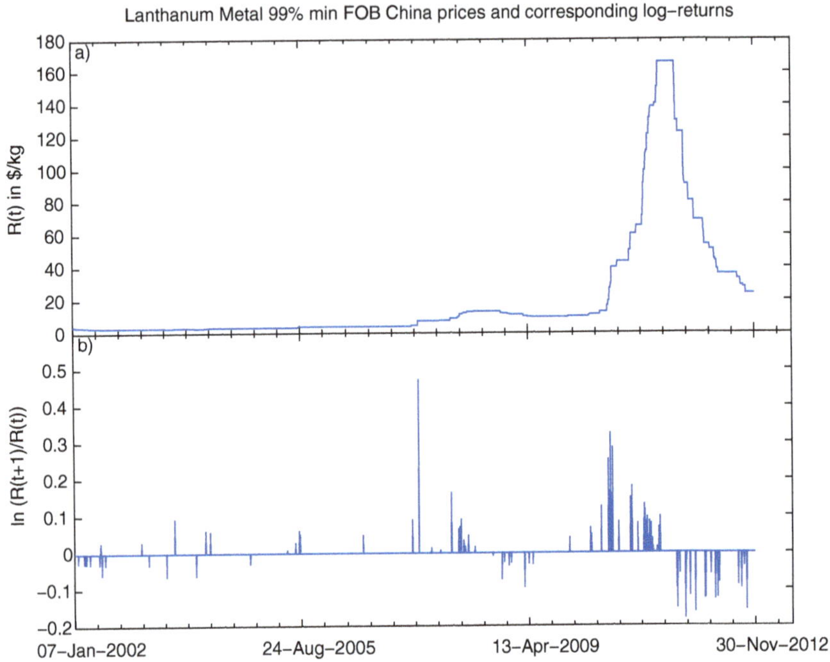

B.3. Estimation results for jump size distributions

Rare earth element/ oxide (REE/REO)	$\hat{\lambda}$	Fitted Distribtution	$\hat{\mu}$	$\hat{\sigma}$	$\hat{\nu}$	p- value ks-test
Cerium Carbonate	0.066	t-location	-0.076	0.026	$\to \infty$*	0.735
Cerium Metal	0.030	t-location	0.001	0.086	4.025	0.287
Cerium Oxide	0.046	t-location	-0.01	0.065	2.479	0.284
Dysprosium Metal	0.045	t-location	0.025	0.04	2.101	0.376
Dysprosium Oxide	0.070	t-location	0.01	0.031	2.076	0.705
Europium Metal	0.025	t-location	-0.012	0.041	1.577	0.635
Europium Oxide	0.049	t-location	0.008	0.039	2.088	0.338
Ferro Dysprosium	0.104	t-location	-0.064	0.061	$\to \infty$*	0.527
Gadolinium Metal	0.020	t-location	0.057	0.157	$\to \infty$*	0.831
Gadolinium Oxide	0.035	t-location	0.037	0.137	34.474	0.618
Lanthanum Metal	0.032	t-location	0.012	0.080	4.578	0.483

Lanthanum Oxide	0.053	t-location	0.000	0.064	2.559	0.637
(99% min)						
Lanthanum Oxide (99.999% min)	0.019	t-location	-0.048	0.023	→ ∞*	0.817
Mischmetal La 35%	0.044	t-location	0.016	0.133	15.984	0.928
Ce 65%						
Mischmetal Low Zinc	0.026	t-location	0.027	0.068	2.084	0.552
and Magnesium						
Neodymium Metal	0.079	t-location	0.007	0.038	3.351	0.135
Neodymium Oxide	0.085	t-location	0.008	0.031	2.363	0.988
Praseodymium Metal	0.055	t-location	0.018	0.040	2.612	0.184
Praseodymium Oxide	0.070	t-location	0.013	0.034	1.929	0.999
Pr-Nd Metal	0.096	t-location	0.006	0.045	3.449	0.188
Pr-Nd Oxide	0.083	t-location	0.002	0.051	3.712	0.694
Samarium Metal	0.021	t-location	0.016	0.105	19.878	0.154
Samarium Oxide	0.019	t-location	-0.005	0.093	2.050	0.382
Terbium Metal	0.038	t-location	0.006	0.047	2.431	0.362
Terbium Oxide	0.050	t-location	0.005	0.055	2.852	0.040
Yttrium Metal	0.019	t-location	0.000	0.093	4.093	0.886
Yttrium Oxide	0.045	t-location	0.023	0.083	3.519	0.686

* Small sample size.

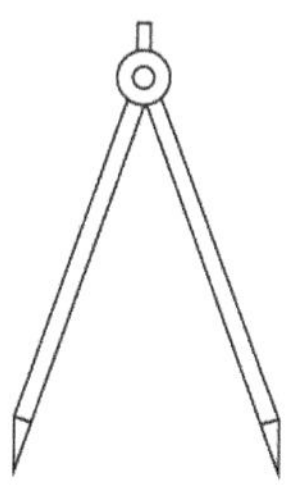

Bibliography

Aase, K. K. (1988). Contingent claims valuation when the security price is a combination of an Ito process and a random point process. *Stochastic Processes and their Applications*, 28(2):185–220.

Ajiferuke, I., Burell, Q., and Tague, J. (1988). Collaborative coefficient: A single measure of the degree of collaboration in research. *Scientometrics*, 14(5):421–433.

Ajodhia, V., Petrov, K., and Scarsi, G. (2003). Benchmarking and its applications. *Zeitschrift für Energiewirtschaft*, 27(4):261–274.

Ariño, A., De la Torre, J., and Ring, P. S. (2001). Relational quality: Managing trust in corporate alliances. *California Management Review*, 44(1):109–131.

Askari, H. and Krichene, N. (2008). Oil price dynamics (2002-2006). *Energy Economics*, 30(5):2134–2153.

Babl, C. (2014). German environmental economics research at a glance: A bibliometric impact analysis. *Zeitschrift für Umweltpolitik und Umweltrecht*, 37(1):1–27.

Babl, C., Schiereck, D., and Flotow, P. (2014). Clean technologies in german economic literature: a bibliometric analysis. *Review of Managerial Science*, 8(1):63–88. Available from: http://dx.doi.org/10.1007/s11846-012-0095-8.

Bastian, M., Heymann, S., and Jacomy, M. (2009). Gephi: An open source software for exploring and manipulating networks. Available from: http://www.aaai.org/ocs/index.php/ICWSM/09/paper/view/154.

Baumgärtner, S. and Quaas, M. (2010). What is sustainability economics? *Ecological Economics*, 69(3):445–450.

Baur, B.; Erhardt, A. (1995). Habitat fragmentation and habitat alterations: principal threats to most animal and plant species. *GAIA*, 4(4):221–226.

Bernhart, W., Schlick, T., Olschewski, I., Thoennes, M., and Garrelfs, J. (2013). Roland Berger Strategy Consultants: E-mobility index for Q3 2013.Available from: http://www.rolandberger.com/media/pdf/ Roland_Berger_E_Mobility_Index_Q3_E_20130905.pdf.

Bertsimas, D., Kogan, L., and Lo, A. W. (1997). Pricing and Hedging Derivative Securities in Incomplete Markets: An *E*-Arbitrage Approach. NBER Working Paper 6250. Available from: www.nber.org/papers/ w6250.

Biewald, A. (2011). Give Virtual Water a Chance! An Attempt to Rehabilitate the Concept. *GAIA*, 20(3):168–170.

Binswanger, M. (1994). Bewirkt wirtschaftlicher Strukturwandel einen Rückgang der Energieintensität? *GAIA*, 3(6):346–352.

Binswanger, M. (1995). Sustainable Development: Utopie in einer wachsenden Wirtschaft? *Zeitschrift für Umweltpolitk und Umweltrecht (ZfU)*, 18(1):1–19.

Black, F. (1976). The pricing of commodity contracts. *Journal of Financial Economics*, 3(1-2):167–179.

Black, F. and Scholes, M. (1973). The pricing of options and corporate liabilities. *Journal of Political Economy*, 81(3):631–654.

BMU (2014). Federal Ministry for the Environment, Nature Conservation, Building and Nuclear safety. Definition der Elektromobilität nach der Bundesregierung. Available from: http://www.erneuerbar-mobil.de/schlagwortverzeichnis/definition-der-elektromobilitaet-nach-der-bundesregierung.

Boehmer, E., Masumeci, J., and Poulsen, A. B. (1991). Event-study methodology under conditions of event-induced variance. *Journal of Financial Economics*, 30(2):253–272.

Bollen, N. P. (1998). Valuing options in regime-switching models. *Journal of Derivatives*, 6:38–49.

Bordons, M. and Gómez, I. (2000). Collaboration networks in science. *The web of knowledge: A festschrift in honor of Eugene Garfield*, pages 197–213.

Braam, R., Moed, H., and van Raan, A. F. J. (1991a). Mapping of science by combined co-citation and word analysis. I. Structural aspects. *Journal of the American Society for Information Science*, 42(4):233–251.

Braam, R., Moed, H., and van Raan, A. F. J. (1991b). Mapping of science by combined co-citation and word analysis. II. Dynamical aspects. *Journal of the American Society for Information Science*, 42(4):252–266.

Brauner, G. (2008). Infrastrukturen der Elektromobilität. *e & i Elektrotechnik und Informationstechnik*, 125(11):382–386. Available from: http://dx.doi.org/10.1007/s00502-008-0585-1.

Brennan, M. J. and Schwartz, E. S. (1985). Evaluating natural ressource investments. *The Journal of Business*, 58(2):135–157.

Broadus, R. N. (1987). Toward a defintion of "bibliometrics". *Scientometrics*, 12(5-6):373–379.

Brooke, J. and Oliver, B. (2005). The source of abnormal returns from strategic alliance announcements. *Pacific-Basin Finance Journal*, 13(2):145–161.

Brooks, C., Lazar, E., and Prokopczuk, M. (2011). Futures basis, scarcity and commodity price volatility: An empirical analysis. *ICMA Centre Discussion Papers in Finance DP2011-20*, pages 1–29.

Brown, S. J. and Warner, J. B. (1985). Using daily stock returns: The case of event studies. *Journal of financial economics*, 14(1):3–31.

Bräuer, W. (2002). Ordnungspolitischer Vergleich von Instrumenten zur Förderung erneuerbarer Energien im deutschen Stromsektor. *Zeitschrift für Umweltpolitk und Umweltrecht (ZfU)*, 25(1):61–103.

Campart, S. and Pfister, E. (2007). Technology, cooperation and stock market value: an event study of new partnership announcements in the biotechnology and pharmaceutical industries. *Economics of Innovation and New Technology*, 16(1):31–49.

Canzler, W., Andreas Knie, A., and Berthold, O. (1993). Das Leitbild Automobil vor der Auflösung? *Zeitschrift für Umweltpoltik und Umweltrecht (ZfU)*, 16(4):407–429.

Castro e Silva, M. and Teixeira, A. A. (2011). A bibliometric account of the evolution of EE in the last two decades: Is ecological economics (becoming) a post-normal science? *Ecological Economics*, 70(5):849–862.

Chan, S. H., Kensinger, J. W., Keown, A. J., and Martin, J. D. (1997). Do strategic alliances create value? *Journal of Financial Economics*, 46(2):199–221.

Cheang, G. H. L. and Chiarella, C. (2008). Hedge Portfolios in Markets with Price Discontinuities. University of Technology Sydney Research Paper No. 218. Available from: http://ssrn.com/abstract=1129063.

Costanza, R., Stern, D., Fisher, B., He, L., and Ma, C. (2004). Influential publications in ecological economics: a citation analysis. *Ecological Economics*, 50(3):261–292.

Cox, J. C. and Ross, S. A. (1976). The valuation of options for alternative stochastic processes. *Journal of Financial Economics*, 3(1):145–166.

Cox, J. C., Ross, S. A., and Rubinstein, M. (1979). Option pricing: A simplified approach. *Journal of Financial Economics*, 7(3):229–263.

Danzon, P. M., Nicholson, S., and Pereira, N. S. (2005). Productivity in pharmaceutical-biotechnology R&D: the role of experience and alliances. *Journal of health economics*, 24(2):317–339.

Das, S., Sen, P. K., and Sengupta, S. (1998). Impact of strategic alliances on firm valuation. *Academy of Management Journal*, 41(1):27–41.

Davidson, R. and MacKinnon, J. G. (1993). *Estimation and inference in econometrics*. Oxford University Press.

Davis, M. H. A. (1998). Option pricing in incomplete markets. In Dempster, M. and Pliska, S., editors, *Mathematics of Derivative Securities*, pages 216–226. Cambridge University Press.

Dempster, A. P., Laird, N. M., and Rubin, D. B. (1977). Maximum likelihood from incomplete data via the em algorithm. *Journal of the Royal Statistical Society. Series B.*, 39(1):1–38.

Die Bundesregierung (2011). Regierungsprogramm Elektromobilität. Available from: http://www.bmbf.de/pubRD/programm_ elektromobilitaet.pdf.

Dombrowsky, W. R. (2002). Flußhochwasser - ein Störfall der Vernunft? *GAIA*, 11(4):310–311.

Dreher, M., Graehl, S., Wietschel, M., and Rentz, O. (2000). Entwicklungstendenzen bei Grünen Angeboten in Deutschland. *Zeitschrift für Energiewirtschaft (ZfE)*, 23(4):191–199.

Drillisch, J. (1999). Quotenregelung für regenerative Stromerzeugung. *Zeitschrift für Energiewirtschaft (ZfE)*, 23(4):251–274.

Duan, J.-C., Popova, I., and Ritchken, P. (2002). Option pricing under regime switching. *Quantitative Finance*, 2:116–132.

Dudenhöffer, F., Bussmann, L., and Dudenhöffer, K. (2012). Elektromobilität braucht intelligente Förderung. *Wirtschaftsdienst*, 92(4):274–279. Available from: http://dx.doi.org/10.1007/s10273-012-1374-3.

Duffie, D. and Richardson, H. R. (1991). Mean-variance hedging in continuous time. *Annals of Applied Probability*, 1(1):1–15.

Eades, P. (1984). A heuristic for graph drawing. *Congressus Numerantium*, 42:149–160.

Egner, H. (2007). Überraschender Zufall oder gelungene wissenschaftliche Kommunikation: Wie kam der Klimawandel in die aktuelle Debatte. *GAIA*, 16(4):250–254.

Eisenbeiß, G. and Goetzberger, A. (2011). Of masterminds and disasters: The history of photovoltaics in Germany. Interview with Gerd Eisenbeiß and Adolf Goetzberger. *GAIA*, 20(4):236–242.

Elliott, R. J. and Kopp, P. E. (1990). Option pricing and hedge portfolios for poisson processes. *Stochastic Analysis Applications*, 8(2):157–167.

Fama, E. F. (1970). Efficient capital markets: A review of theory and empirical work. *The Journal of Finance*, 25(2):383–417. Available from: http://www.jstor.org/stable/2325486.

Fama, E. F. (1998). Market efficiency, long-term returns, and behavioral finance. *Journal of Financial Economics*, 49(3):283–306. Available from: http://www.sciencedirect.com/science/article/pii/ S0304405X98000269.

Fama, E. F. and French, K. R. (1987). Commodity futures prices: Some evidence on forecast power, premiums, and the theory of storage. *The Journal of Business*, 60(1):55–73.

Fama, E. F. and French, K. R. (1988). Business cycles and the behaviour of metals prices. *Journal of Finance*, 43(5):1075–1093.

Frondel, M. and Peters, J. (2007). Biodiesel: Eine teure Klimaschutzoption. *Zeitschrift für Umweltpolitk und Umweltrecht (ZfU)*, 30(2):233–251.

Frondel, M., Ritter, N., Moore, N., and Schmidt, C. (2011). Die Kosten des Klimaschutzes am Beispiel der Strompreise für private Haushalte. *Zeitschrift für Energiewirtschaft*, 35(3):195–207. Available from: http: //dx.doi.org/10.1007/s12398-011-0059-8.

Fruchterman, T. M. J. and Reingold, E. M. (1991). Graph drawing by forcedirected placement. *Software-practice and Experience*, 21(11):1129–1164.

Fuller, K., Netter, J., and Stegemoller, M. (2002). What do returns to acquiring firms tell us? Evidence from firms that make many acquisitions. *The Journal of Finance*, 57(4):1763–1793.

Garfield, E. (1979). Is citation analysis a legitimate evaluation tool? *Scientometrics*, 1(4):359–375.

Garfield, E. (2001). From bibliographic coupling to co-citation analysis via algorithmic historio-bibliography: A citationist's tribute to Belver C. Griffith, Lazerow Lecture presented at Drexel University, Philadelphia, PA. November 27, 2001. Available from: http://www.garfield. library.upenn.edu/papers/drexelbelvergriffith92001.pdf.

Gawel, E. and Korte, K. (2012). Verteilungswirkungen des EEG: Wird die Energiewende ungerecht organisiert? *Zeitschrift für Umweltrecht*, 23(9):457–458.

Gerpott, T. J. (2009). Einflüsse anbieterbezogener Einstellungen von Privatkunden auf deren Preisbereitschaft. *Die Betriebswirtschaft (DBW)*, 69(6):679–700.

Gesang, B. and Wiesenthal, T. (2003). Windenergie ausbauen? - Eine technik-ethische Evaluation der Kontroverse zwischen Natur- und Umweltschützern. *Zeitschrift für Umweltpolitik und Umweltrecht (ZfU)*, 26(2):169–186.

Gethmann, C. F. and Mittelstraß, J. (1992). Maße für die Umwelt. *GAIA*, 1(1):16–25.

Geuna, A. and Martin, B. R. (2003). University research evaluation and funding: An international comparison. *Minerva*, 41(4):277–304.

Gibson, R. and Schwartz, E. S. (1990). Stochastic convenience yield and the pricing of oil contingent claims. *The Journal of Finance*, 45(3):959–976.

Gisler, P. and Pohl, C. (2000). Die Züricher Solarstrombörse - was sich aus einer Erfolgsstory lernen läßt. *GAIA*, 9(2):128–131.

Gmür, M. (2003). Co-citation analysis and the search for invisible colleges: A methodological evaluation. *Scientometrics*, 57(1):27–57.

Graehl, S., Dreher, M., Wietschel, M., and Rentz, O. (2001). Eine Analyse des Marktes für Grüne Angebote in Deutschland. *Zeitschrift für Energiewirtschaft (ZfE)*, 25(4):221–230.

Green, P. J. (1995). Reversible jump markov chain monte carlo computation and bayesian model determination. *Biometrika*, 82(4):711–732.

Hamers, L., Hemeryck, Y., Herweyers, G., Janssen, M., Keters, H., Rousseau, R., and Vanhoutte, A. (1989). Similarity Measures In Scientometric Research: The Jaccard Index Versus Salton's Cosine Formula. *Information Processing & Manogement*, 25(3):315–318.

Hamilton, J. D. (1989). A new approach to the economic analysis of non-stationary time series and the business cycle. *Econometrica*, 57(2):357–384.

Hanselka, H. and Jöckel, M. (2010). Elektromobilität: Elemente, Herausforderungen, Potenziale. In Hüttl, R. F., Pischetsrieder, B., and Spath, D., editors, *Elektromobilität: Potenziale und Wissenschaftlich-Technische Herausforderungen*, pages 21–38. Springer, Berlin, Heidelberg.

Hansjürgens, B. and Fromm, O. (1994). Erfolgsbedingungen von Zertifikatelösungen in der Umweltpolitik - am Beispiel der Novelle des US-Clean Air Act von 1990. *Zeitschrift für Umweltpolitik und Umweltrecht*, 17(4):473–505.

Harrison, J. M. and Pliska, S. R. (1981). Martingales and stochastic integrals in the theory of continous trading. *Stochastic processes and their applications*, 11(3):215–260.

Havemann, F. (2009). Einführung in die Bibliometrie. Gesellschaft für Wissenschaftsforschung, Berlin. Available from: http://www. wissenschaftsforschung.de/Havemann2009Bibliometrie.pdf.

Hemmelskamp, J. (1999). Der Einfluss der Umweltpolitik auf das Innovationsverhalten. *Zeitschrift für Umweltpolitk und Umweltrecht (ZfU)*, 22(1):33–66.

Henseling, K. O. (2005). Nachhaltigkeit und Auto-Mobilität. *GAIA*, 14(4): 299–306.

Heymann, E., Koppel, O., and Puls, T. (2013). *Evolution statt Revolution: die Zukunft der Elektromobilität*. IW-Analysen. Inst. der Dt. Wirtschaft, Köln.

Hofstetter, P. (1992). Weshalb Öko-Kühlschrank? FCKW-, Energie- und Entsorgungsproblematik. *GAIA*, 1(6):350–360.

Homburg, C., Giering, A., and Hentschel, F. (1999). Der Zusammenhang zwischen Kundenzufriedenheit und Kundenbindung. *Die Betriebswirtschaft (DBW)*, 59(2):174–195.

Hubalek, F., Kallsen, J., and Krawczyk, L. (2006). Variance-optimal hedging for processes with stationary independent increments. *The Annals of Applied Probability*, 16(2):853–885.

Huckestein, B. (1993). Umweltlizenzen - Anwendungsbedingungen einer ökonomisch effizienten Umweltpolitik durch Mengensteuerung. *Zeitschrift für Umweltpolitik und Umweltrecht*, 16(1):1–29.

Humphries, M. (2012). Rare Earth Elements: The Global Supply Chain. Congressional Research Service. Available from: http://www.fas.org/sgp/crs/natsec/R41347.pdf.

Jaeger, J. and Holderegger, R. (2005). Schwellenwerte der Landschaftszerschneidung. *GAIA*, 14(2):113–118.

Jaeger, J. and Scheringer, M. (2006). Einführung: Warum trägt die Umweltforschung nicht stärker zur Lösung von Umweltproblemen bei? *GAIA*, 15(1):20–23.

Jeske, T. and von Hirschhausen, C.. (2005). Offshore Windenergie: Studie zur Rentabilität von Offshore-Windparks in der Deutschen Nord- und Ostsee. *Zeitschrift für Energiewirtschaft (ZfE)*, 29(1):3–11.

Jänicke, M. (1990). Erfolgsbedingungen von Umweltpolitik im internationalen Vergleich. *ZfU*, 13:213–232.

Johnson, L. L. (1960). The theory of hedging and speculation in commodity futures. *The Review of Economic Studies*, 27(3):139–151.

Kaldor, N. (1939). Speculation and economic stability. *Review of economic studies*, 7:1–27.

Kale, P., Singh, H., and Perlmutter, H. (2000). Learning and protection of proprietary assets in strategic alliances: building relational capital. *Strategic Management Journal*, 21(3):217–237.

Kannen, A. and Burkhard, B. (2009). Integrated assessment of coastal and marine changes using the example of offshore wind farms: the coastal futures approach. *GAIA*, 18(3):229–238.

Kasperk, G. and Drauz, R. (2013). Geschäftsmodelle entlang der elektromobilen Wertschöpfungskette. In Kampker, A., Vallée, D., and Schnettler, A., editors, *Elektromobilität: Grundlagen einer Zukunftstechnologie*, pages 103–148. Springer, Heidelberg, Berlin.

Kessler, M. M. (1963). Bibliographic coupling between scientific papers. *American Documentation*, 14(1):10–25.

Knie, A. and Berthold, O. (1996). Ceteris Paribus-Syndrom in der Mobilitätspolitik? Tatsächliche Nutzungsprofile von elektrischen Straßenfahrzeugen. *Zeitschrift für Umweltpolitk und Umweltrecht (ZfU)*, 19(1):75–98.

Koh, J. and Venkatraman, N. (1991). Joint venture formations and stock market reactions: An assessment in the information technology sector. *Academy of Management Journal*, 34(4):869–892.

Kohler, H. (2010). Herausforderungen im Bereich Fahrzeugkonzepte und Elektrische Antriebssysteme. In Hüttl, R. F., Pischetsrieder, B., and Spath, D., editors, *Elektromobilität: Potenziale und Wissenschaftlich-Technische Herausforderungen*, pages 75–84. Springer, Heidelberg, Berlin.

Kuhn, T. S. (1962). *The structure of scientific revolutions*. University of Chicago press.

Leitinger, C. and Litzlbauer, M. (2011). Netzintegration und Ladestrategien der Elektromobilität. *e & i Elektrotechnik und Informationstechnik*, 128(1-2):10–15. Available from: http://dx.doi.org/10.1007/ s00502-011-0800-3.

Leydesdorf, L. and Shin, J. C. (2010). How to evaluate universities in terms of their relative citation impacts: Fractional counting of citations and the normalization of differences among disciplines. Available from: arXiv:1010.2465.

Lienkamp, M. (2012). *Elektromobilität: Hype oder Revolution? Die Endlichkeit von Ölvorräten und der Klimawandel*. Springer, Heidelberg, Berlin.

Long, K. R., Gosen, B. S. V., Foley, N. K., and Cordier, D. (2010). The Principal Rare Earth Elements Deposits of the United States - A Summary of Domestic

Deposits and a Global Perspective. U.S. Geological Survey Scientific Investigations Report 2010–5220. Available from: http://pubs.usgs.gov/sir/2010/5220.

Lotka, A. J. (1926). The frequency distribution of scientific productivity. *Journal of the Washington Academy of Sciences*, 16(12):317–323.

Luhmann, H.-J. (2008). CCS: Ein Beitrag zur Bekämpfung des Klimawandels - oder zu dessen Steigerung? *Zeitschrift für Umweltpolitk und Umweltrecht (ZfU)*, 31(1):141–154.

Ma, C. and Stern, D. I. (2006). Environmental and ecological economics: A citation analysis. *Ecological Economics*, 58(3):491–506.

MacKinlay, A. C. (1997). Event studies in economics and finance. *Journal of economic literature*, 35(1):13–39.

Madlener, R., Siegers, L., and Bendig, S. (2009). Risikomanagement und -controlling bei Offshore-Windenergieanlagen. *Zeitschrift für Energiewirtschaft (ZfE)*, 33(2):135–146.

Maier, J. (2010). Neue Materialien für die Elektromobilität: Potenzial und Stellschrauben. In *Elektromobilität*, pages 63–74. Springer, Heidelberg, Berlin.

Marshakova, I. (1973). System of document connections based on references. *Nauchno-Tekhnicheskaya Informatsiya Seriya 2-Informatsionnye Protsessy I Sistemy*, 2(6):3–8.

Marshall, J. and Toffel, M. (2005). Framing the elusive concept of sustainability: A sustainability hierarchy. *Environmental & Scientific Technology*, 39(3):673–682.

McCarthy, J. L. (1982). Metadata management for large statistical databases. In *Proceedings of the 8th International Conference on Very Large Data Bases*, pages 234–243. Morgan Kaufmann Publishers Inc.

McConnell, J. and Nantell, T. J. (1985). Corporate combinations and common stock returns: The case of joint ventures. *The Journal of Finance*, 40(2):519–536.

Meisel, R., Pforte, R., and Fichtner, W. (2009). Development of onshore wind energy utilisation in Germany. *Zeitschrift für Energiewirtschaft. (ZfE)*, 33(1):3–13.

Menges, R. and Traub, S. (2008). Staat versus Markt: Konsumentenpräferenzen und die Förderung erneuerbarer Energien. *Zeitschrift für Energiewirtschaft (ZfE)*, 32(4):262–270.

Merton, R. C. (1976). Option Pricing When Underlying Stock Returns Are Discontinuous. *Journal of Financial Economics*, 3(1):125–144.

Merton, R. C. (1998). Applications of option-pricing theory: Twenty-five years later. *The American Economic Review*, 88(3):323–349.

Moody, J., Bender-deMoll, S., and McFarland, D. A. (2006). Dynamic Network Visualization. *American Journal of Sociology*, 110(5):1206–1241.

Morche, D., Schmitt, F., Genuit, K., Elsen, O., Kampker, A., and Friedrich, B. (2013). Fahrzeugkonzeption für die Elektromobilität. In *Elektromobilität*, pages 149–234. Springer.

Müller, H. (2012). Zitationen als Grundlage von Forschungsleistungsrankings - Konzeptionelle Überlegungen am Beispiel der Betriebswirtschaftslehre. *Beiträge zur Hochschulforschung*, 34(2):68–92.

Nationale Plattform Elektromobilität (2012). Fortschrittsbericht der Nationalen Plattform Elektromobilität (Dritter Bericht). Available from: http://www.bmub.bund.de/fileadmin/bmu-import/files/ pdfs/allgemein/application/pdf/bericht_emob_3_bf.pdf.

Nieto, M. J. and Santamar'ıa, L. (2010). Technological collaboration: Bridging the innovation gap between small and large firms. *Journal of Small Business Management*, 48(1):44–69.

Ohl, C. and Eichhorn, M. (2008). Nachhaltige Landnutzung im Kontext der Windenergie - Rationierung von Flächen als Antwort auf die energiepolitischen Ziele der Klimapolitik. *Zeitschrift für Umweltpolitk und Umweltrecht (ZfU)*, 31(4):517–540.

Okubo, Y. (1997). Bibliometric indicators and analysis of research systems: Methods and examples. *OECD Science, Technology and Industry Working Papers, 1997/01, OECD Publishing.*

Osareh, F. (1996). Bibliometrics, citation analysis and co-citation analysis: A review of literature i. *Libri*, 46(3):149–158.

Page, F. H. and Sanders, A. B. (1986). A General Derivation of the Jump Process Option Pricing Formula. *The Journal of Financial and Quantitative Analysis*, 21(4):437–446.

Papp, J. F., Bray, E. L., Edelstein, D. L., Fenton, M. D., Guberman, D. E., Hedrick, J. B., Jorgenson, J. D., Kuck, P. H., Shedd, K. B., and Tolcin, A. C. (2008). Factors that influence the price of Al, Cd, Co, Cu, Fe, Ni, Pb, Rare Earth Elements, and Zn. U.S. Geological Survey. Open-File Report 2008–1356. Available from: http://pubs.usgs.gov/of/2008/1356/pdf/ofr2008-1356.pdf.

Potter, W. G. (1988). "Of Making Many Books There Is No End": Bibliometrics and Libraries. *Journal of Academic Librarianship*, 14(4):238a–38c.

PricewaterhouseCoopers (PwC) (2011). Minerals and metals scarcity in manufacturing: The ticking timebomb. Available from: http://www.pwc.com/en_GX/gx/sustainability/research-insights/assets/impact-of-minerals-metals-scarcity-on- business.pdf.

Primas, H. (1992). Umdenken in der Naturwissenschaft. *GAIA*, 1(1):5–15.

Proff, H. and Proff, H. (2013). *Dynamisches Automobilmanagement: Strategien für Hersteller und Zulieferer im internationalen Wettbewerb*. Springer, Heidelberg, Berlin, second edition.

Pütz, T. and Schmied, H. (1995). Auswirkungen von Technologiepolitik auf die Entwicklung großtechnischer Systeme - Vergleichende Studie zwischen Deutschland und Frankreich. *Schmalenbachs Zeitschrift für betriebswirtschaftliche Forschung (ZfbF)*, 46(9):857–880.

Radke, V. (1996). Ökonomische Aspekte nachhaltiger Technologie. *Zeitschrift für Umweltpolitik und Umweltrecht (ZfU)*, 19(1):109–128.

Reller, A., Bublies, T., Staudinger, T., Oswald, I., Meißner, S., and Allen, M. (2009). The mobile phone: Powerful communicator and potential metal dissipator. *GAIA*, 18(2):127–135.

Riechmann, C. (1995). Price-Cap Regulierung. *Zeitschrift für Energiewirtschaft*, 2(95):157–167.

Rommel, K. and Meyerhoff, J. (2009). Empirische Analyse des Wechselverhaltens von Stromkunden. Was hält Stromkunden davon ab, zu Ökostromanbietern zu Wechseln? *Zeitschrift für Energiewirtschaft (ZfE)*, 33(1):74–82.

Routledge, B. R., Seppi, D. J., and Spatt, C. S. (2000). Equilibrium forward curves for commodities. *The Journal of Finance*, 55(3):1297–1338.

Russell, J. and Rousseau, R. (2002). Bibliometrics and institutional evaluation. *Encyclopedia of Life Support Systems (EOLSS)*, pages 42–64.

Rydén, T. (2008). EM versus Markov chain Monte Carlo for Estimation of Hidden Markov Models: A Computational Perspective. *Bayesian Analysis*, 3(4):659–688.

Sachse, M. and Bach, W. (1994). CO2-Vermeidung durch Solarenergie: Potentiale und Kosten am Beispiel Münsters. *Zeitschrift für Energiewirtschaft (ZfE)*, 18(4):227–238.

Schäffer, U., Binder, C., and Gmür, M. (2006). Struktur und Entwicklung der Controllingforschung - Eine Zitations- und Kozitationsanalyse von Controllingbeiträgen in deutschsprachigen wissenschaftlichen Zeitschriften von 1970 bis 2003. *Zeitschrift für Betriebswirtschaft (ZfB)*, 76(4):395–440.

Schäffer, U., Nevries, P., Fikus, C., and Meyer, M. (2011). Is Finance Research a "Normal Science"? A Bibliometric Study of the Structure and Development of Finance Research from 1988 to 2007. *Schmalenbach Business Review (sbr)*, 63(2):189–225.

Schanze, J. (2002). Nach der Elbeflut 2002: Die gesellschaftliche Risikovorsorge bedarf einer transdisziplinen Hochwasserforschung. *GAIA*, 11(4):247–254.

Schidler, S. and Sotoudeh, M. (2005). Bioraffinerie kombiniert mit Biogas. *GAIA*, 14(4):354–356.

Schrader, U. and Hennig-Thurau, T. (2009). Method, results, and implications of the german academic association for business research's journal ranking. *BuR - Business Research*, 2(2):180–204.

Schweizer, M. (1992). Mean-variance hedging for general claims. *Annals of applied probability*, 2(1):171–179.

Schweizer, M. (1995). Variance-Optimal Hedging in Discrete Time. *Mathematics of Operations Research*, 20(1):1–31.

Shreve, S. E. (2004). *Stochastic Calculus for Finance II: Continuous-Time Models*. Springer.

Siegel, S. (1956). *Nonparametric statistics for the behavioral sciences*. McGraw-Hill.

Small, H. (1973). Co-citation in the scientific literature: A new measure of the relationship between two documents. *Journal of the American Society for Information Science*, 24(4):265–269.

Small, H. G. and Griffith, B. (1974). The structure of scientific literature I: Identifying and graphing specialities. *Science Studies*, 4:17–40.

Smith, L. (2012). *The Rare Earth Market: A Monthly Analysis for the Rare Earth Industry. August 2012*. Metal-Pages.

Solow, R. M. (1974). The economics of resources or the resources of economics. *The American Economic Review*, 64(2):1–14.

Spertus, E., Sahami, M., and Buyukkokten, O. (2005). Evaluating similarity measures: a large-scale study in the orkut social network. In *Proceedings of the eleventh ACM SIGKDD international conference on Knowledge discovery in data mining*, pages 678–684. ACM.

Subramanyam, K. (1983). Bibliometric studies of research collaboration: A review. *Journal of Information Science*, 6(1):33–38.

The Cleantech Group (2014). Cleantech definition: Clean is more than green. Available from: http://www.cleantech.com/what-is-cleantech/.

ThomsonReuters (2013). Thomson Reuter indices: Global Equity Index Methodology. April 2013. Available from: http://thomsonreuters. com/products/financial-risk/01_257/global-equity-index- methodology-april-2013.pdf.

Tietenberg, T. H. (1980). Transferable discharge permits and the control of stationary source air pollution: a survey and synthesis. *Zeitschrift für Umweltpolitik*, pages 477–508.

Tse, P.-K. (2011). China's Rare-Earth Industry. U.S. Geological Survey. Open–File Report 2011–1042. Available from: http://pubs.usgs. gov/of/2011/1042/of2011-1042.pdf.

USGS (2012). US Geological Survey. Mineral Commodity Summary: Rare Earths 2012. Available from: http://minerals.usgs.gov/minerals/ pubs/commodity/rare_earths/mcs-2012-raree.pdf.

van den Bergh, J. C. (2001). Ecological economics: themes, approaches, and differences with environmental economics. *Regional Environmental Change*, 2(1):13–23.

Van Raan, A. F. (2005). Fatal attraction: Conceptual and methodological problems in the ranking of universities by bibliometric methods. *Scientometrics*, 62(1):133–143.

VDA (2013). Verband der Automobilindustrie: Grafik Antriebstechnologien. Available from: http://www.vda.de/de/downloads/959/.

Verbeek, A., Debackere, K., Luwel, M., and Zimmermann, E. (2002). Measuring progress and evolution in science and technology - I: The multiple uses of bibliometric indicators. *International Journal of Management Reviews*, 4(2):179–211.

Wagner, G. R. and Matten, D. (1992). Die Erstellung innovativer technischer Großsysteme im Spannungsfeld zwischen ökonomischen und ökologischen Risiken - diskutiert am Fall Transrapid. *Die Betriebswirtschaft (DBW)*, 52:359–383.

Wallentowitz, H., Freialdenhoven, A., and Olschewski, I. (2008). *Strategien in der Automobilindustrie: Technologietrends und Marktentwicklungen*. Springer, Heidelberg, Berlin.

Wallentowitz, H., Freialdenhoven, A., and Olschewski, I. (2010). *Strategien zur Elektrifizierung des Antriebstranges: Technologien, Märkte und Implikationen*. Springer, Heidelberg, Berlin.

Weitzman, M. L. (2013). Tail-hedge discounting and the social cost of carbon. *Journal of Economic Literature*, 51(3):873–882.

Weller, T. (1998). Green Pricing: kundenorientierte Angebote der Elektrizitätswirtschaft. *Zeitschrift für Energiewirtschaft (ZfE)*, 22(1):58–70.

Wenzel, B., Ohlhorst, D., and Bruns, E. (2009). Geothermische Stromerzeugung in Deutschland - Stiefkind oder schlafender Riese? *Zeitschrift für Energiewirtschaft (ZfE)*, 33(1):23–30.

White, H. (1980). A heteroskedasticity-consistent covariance matrix estimator and a direct test for heteroskedasticity. *Econometrica: Journal of the Econometric Society*, 48(4):817–838.

White, H. D. and Griffith, B. C. (1981). Author cocitation: A literature measure of intellectual structure. *Journal of the American Society for Information Science*, 32(3):163–171.

Wilcoxon, F. (1945). Individual comparisons by ranking methods. *Biometrics bulletin*, 1(6):80–83.

Working, H. (1948). Theory of the inverse carrying charge in futures markets. *Journal of Farm Economics*, 30(1):1–28.

Working, H. (1949). The theory of the price of storage. *American Economic Review*, 39(6):1254–1262.

WTO (2012). World Trade Organisation. China — Measures Related to the Exportation of Rare Earths, Tungsten and Molybdenum. Dispute Settlements DS 431, DS 432, DS 433. Available from: http://www. wto.org/english/tratop_e/dispu_e/dispu_status_e.htm.

Yay, M. (2010). *Elektromobilität: theoretische Grundlagen, Herausforderungen sowie Chancen und Risiken der Elektromobilität, diskutiert an den Umsetzungsmöglichkeiten in die Praxis*. Peter Lang, Frankfurt.

Ziehm, C. (2012). Ohne Solidarität keine Energiewende. *Zeitschrift für Umweltrecht*, 23(11):585–586.

Zinnbauer, M. (2001). Cross-Selling-Potenziale bei Energieversorgungsunternehmen durch Bundling. *Zeitschrift für Energiewirtschaft (ZfE)*, 25(4): 243–252.

Finanzmärkte und Klimawandel

Herausgegeben von Dirk Schiereck und Paschen von Flotow

Band 1 Christian Babl / Paschen von Flotow / Dirk Schiereck (Hrsg.): Projektrisiken und Finanzierungsstrukturen bei Investitionen in erneuerbare Energien. 2011.

Band 2 Christoph Ettenhuber: Financing Corporate Growth in the Renewable Energy Industry. 2013.

Band 3 Anette von Ahsen / Robert Fraunhoffer / Dirk Schiereck (Hrsg.): Wellenbrecher auf dem Weg zur Energiewende? Zur Attraktivität von Energiespeicherung, nachhaltiger Erzeugung und Verbrauchersteuerung. 2013.

Band 4 Christian Friebe: Diffusion of Renewable Energy Technologies. Private Sector Perspectives on Emerging Markets. 2014.

Band 5 Paschen von Flotow / Dirk Schiereck / Julian Trillig (Hrsg.): Energietransformation, dezentrale Erzeugungsprobleme und Finanzierung der Solarindustrie. 2014.

Band 6 Christian Babl: E-Mobility and Related Clean Technologies from an Empirical Corporate Finance Perspective. State of Economic Research, Sourcing Risks, and Capital Market Perception. 2015.

www.peterlang.com

www.ingramcontent.com/pod-product-compliance
Ingram Content Group UK Ltd.
Pitfield, Milton Keynes, MK11 3LW, UK
UKHW020641041026
15590UKWH00001B/22

9 783631 661390